Springer Theses

Recognizing Outstanding Ph.D. Research

Aims and Scope

The series "Springer Theses" brings together a selection of the very best Ph.D. theses from around the world and across the physical sciences. Nominated and endorsed by two recognized specialists, each published volume has been selected for its scientific excellence and the high impact of its contents for the pertinent field of research. For greater accessibility to non-specialists, the published versions include an extended introduction, as well as a foreword by the student's supervisor explaining the special relevance of the work for the field. As a whole, the series will provide a valuable resource both for newcomers to the research fields described, and for other scientists seeking detailed background information on special questions. Finally, it provides an accredited documentation of the valuable contributions made by today's younger generation of scientists.

Theses are accepted into the series by invited nomination only and must fulfill all of the following criteria

- They must be written in good English.
- The topic should fall within the confines of Chemistry, Physics, Earth Sciences, Engineering and related interdisciplinary fields such as Materials, Nanoscience, Chemical Engineering, Complex Systems and Biophysics.
- The work reported in the thesis must represent a significant scientific advance.
- If the thesis includes previously published material, permission to reproduce this must be gained from the respective copyright holder.
- They must have been examined and passed during the 12 months prior to nomination.
- Each thesis should include a foreword by the supervisor outlining the significance of its content.
- The theses should have a clearly defined structure including an introduction accessible to scientists not expert in that particular field.

More information about this series at http://www.springer.com/series/8790

Asmus Ougaard Dohn

Transient Changes in Molecular Geometries and How to Model Them

Simulating Chemical Reactions of Metal Complexes in Solution to Explore Dynamics, Solvation, Coherence, and the Link to Experiment

Doctoral Thesis accepted by
the Technical University of Denmark, Denmark

Springer

Author
Dr. Asmus Ougaard Dohn
Department of Chemistry
Technical University of Denmark
Kongens Lyngby
Denmark

Supervisors
Prof. Klaus B. Møller
Department of Chemistry
Technical University of Denmark
Kongens Lyngby
Denmark

Prof. Niels E. Henriksen
Department of Chemistry
Technical University of Denmark
Kongens Lyngby
Denmark

ISSN 2190-5053 ISSN 2190-5061 (electronic)
Springer Theses
ISBN 978-3-319-38683-6 ISBN 978-3-319-18747-1 (eBook)
DOI 10.1007/978-3-319-18747-1

Springer Cham Heidelberg New York Dordrecht London

Softcover reprint of the hardcover 1st edition 2015

Printed on acid-free paper

Springer International Publishing AG Switzerland is part of Springer Science+Business Media (www.springer.com)

Part of this thesis has been published in the following journal articles:

The articles listed below were partially or fully prepared during the period of the Ph.D. program. Parts of the content in this thesis are based upon publications IV, VI, VII, and I, while results from V and III are drawn upon when needed.

The articles are listed chronologically. I is still in preparation, while the rest are published.

(I) Haldrup, M.K., Gawelda, W., Abela, R., Alonso-Mori, R., Bergmann, U., Bordage, A., Cammarata, M., Canton, S. Dohn, A.O., Brandt van Driel, T., Fritz, D., Gaffney, K., Galler, A., Glatzel, P., Harlang, T., Kjær, K., Lemke, H., Møller, K.B., Zoltan, N., Pápai, M., Sas, N., Uhlig, J., Zhu, D., Vankó, G., Sundström, V. & Nielsen, M.M. & Bressler, C., 'Observing Solvation Dynamics with Simultaneous Femtosecond X-ray Emission Spectroscopy and X-ray Scattering', *In preparation.*

(II) Vankó, G., Bordage, A., Pápai, M., Haldrup, K., Glatzel, P., March, A.M., Doumy, G., Britz, A., Galler, A., Assefa, T., Cabaret D., Juhin, A., Brandt van Driel, T., Kjær, K., Dohn, A.O., Møller, K.B., Lemke, H.T., Gallo, E., Rovezzi, M., Németh, Z., Rosályi, E., Rozgonyi, T., Uhlig, J., Sundström, V., Nielsen, M.M., Young, L., Southworth, S.H., Bressler, C & Wojciech, G., (2015) 'Detailed Characterization of a Nanosecond-Lived Excited State: X-ray and Theoretical Investigation of the Quintet State in Photoexcited $[Fe(terpy)_2]^{2+}$', *Journal of Physical Chemistry C*, just accepted, DOI: 10.1021/acs.jpcc.5b00557

(III) Canton, S.E., Kjær, K.S., Vankó, G., Brandt van Driel, T., Adachi, S., Bordage, A., Bressler, C., Chabera, P., Christensen, M., Dohn, A.O., Galler, A., Gawelda, W., Gosztola, D., Haldrup, K., Harlang, T., Liu, Y., Møller, K.B., Németh, Z., Nozawa, S., Pápai, M., Sato, T., Sato., T., Suarez-Alcantara, K., Togashi, T., Tono, K., Uhlig, J., Vithanage, D.A., Wärnmark, K., Yabashi, M., Zhang, J., Sundström, V. & Nielsen, M.M., (2015) 'Visualizing the Nonequilibrium Dynamics of Photoinduced Intramolecular Electron Transfer with Femtosecond X-ray Pulses', *Nature Communications*, vol 6, 6359

(IV) Dohn, A.O., Jónsson, E.Ö., Kjær, K.S., Brandt van Driel, T., Nielsen, M. M., Jacobsen, K.W., Henriksen, N.E. & Møller, K.B. (2014). 'Direct Dynamics Studies of a Binuclear Metal Complex in Solution: The Interplay Between Vibrational Relaxation, Coherence, and Solvent Effects'*Journal of Physical Chemistry Letters*, vol 5, pp. 2414–2418.

(V) Canton, S.E., Zhang, J., Brandt van Driel, T., Kjær, K.S., Haldrup, M.K., Chabera, P., Harlang, T., Suarez-Alcantara, K., Liu, Y., Pérez, J-. Bordage, A., Pápai, M., Vankó, G., Jennings, G., Kurtz, C.A., Rovezzi, M., Glatzel, P., Smolentsev, G., Uhlig, J., Dohn, A.O., Christensen, M., Galler, A., Gawelda, W., Bressler, C., Lemke, H.T., Møller, K.B., Nielsen, M.M., Lomoth, R., Wärnmark, K., & Sundström, V. (2013) 'Toward Highlighting the Ultrafast Electron Transfer Dynamics at the Optically Dark Sites of

Photocatalysts'*Journal of Physical Chemistry Letters*, vol 4, pp. 1972–1976.

(VI) Dohn, A.O., Møller, K.B., & Sauer, P.A. (2013), 'Optimizing the Structure of Tetracyanoplatinate (II): A Comparison of Relativistic Density Functional Theory Methods'*Current Inorganic Chemistry*, vol 3, pp. 213–219.

(VII) Haldrup, M.K., Vankó, G., Gawelda, W., Galler, A., Doumy, G., March, A. M., Kanter, E.P., Bordage, A., Dohn, A.O., Brandt van Driel, T., Kjær, K. S., Lemke, H.T., Canton, S.E., Uhlig, J., Sundström, V., Young, L., Southworth, S.H., Nielsen, M.M. & Bressler, C. (2012), 'Guest-Host Interactions Investigated by Time-Resolved X-ray Spectroscopies and Scattering at MHz Rates: Solvation Dynamics and Photoinduced Spin Transition in Aqueous $Fe(bipy)_3^{2+}$'*Journal of Physical Chemistry A*, vol. 116, pp. 9878–9887.

Dedicated to the memory of Mogens Ougaard —who sparked my interest in the wonders of nature by sending me postcards addressed as the marvellous bacteria living in my stomach.

For those who know
It's time to leave the house
And go back to the field
Find your strength in the sound
And make your transition

—Mad Mike, Underground Resistance,
Detroit

Supervisors' Foreword

Structure and change are central concepts in chemistry. In recent years it has become possible to detect the latter at the molecular level where atoms move at very high speeds (equivalent to several km/s). Uncovering these dynamics is central for our understanding of chemical reactions at the fundamental level. This topic is at the heart of femtochemistry, where dynamical nonequilibrium structures are detected while chemical reactions take place. Most recently, large-scale international facilities—so-called X-ray free-electron lasers XFEL—are being built around the world. With these facilities it will become possible to record snapshots of chemical reactions—in the form of instantaneous X-ray scattering signals—at the femtosecond timescale. In parallel with this development of instrumentation, theory and modeling need to be developed in order to interpret all the new experimental data.

To that end, the thesis by Asmus O. Dohn *Transient Changes in Molecular Geometries and How to Model Them* describes the modeling of the dynamics of complex photochemical reactions in the solution phase. The thesis describes in particular the implementation of a highly efficient QM/MM Direct Dynamics approach with application to the photochemistry of transition metal complexes in solution and analyzes various aspects of energy dissipation via dynamic changes in molecular structure, vibrational modes, and solvation. The agreement between theory and experiment is impressive and provides convincing validation of the method, while also providing significant new insight into the chemical dynamics and molecular determinants of the experimental data. Besides providing intriguing information about the solute–solvent dynamics, this type of modeling will also play an important role in the fast growing field of femtosecond X-ray science.

Kongens Lyngby, Denmark
March 2015

Prof. Klaus B. Møller
Prof. Niels E. Henriksen

Abstract

Light-induced chemical processes are accompanied by molecular motion on the femtosecond time scale. Uncovering this dynamical motion is central to understanding the chemical reaction on a fundamental level. This thesis focuses on the aspects of excess excitation energy dissipation via dynamic changes in molecular structure, vibrations and solvation.

In this thesis, we employ our recently developed Quantum-/Molecular-Mechanical Direct Dynamics method to do simulations of transition metal complexes in solution, to uncover their energy dissipation channels, and how they are affected by the solvent. The simulations have also served as benchmarks on this newly developed implementation.

First, we establish that the chosen model provides a trustworthy description of the systems; since transition metals are heavier than purely organic systems, we test a range of approximations to relativistic quantum mechanic descriptions, to ascertain the accuracy of the quantum model in the Direct Dynamics simulations. We then test—and improve—the framework for calculating the experimental X-ray Diffuse Scattering Difference signal from (any kind of) Molecular Dynamics (MD) simulations. Comparisons of purely classical MD simulations to literature Direct Dynamics simulations delineate the boundaries for the force field approximation: Classical MD provides a solvent shell response sufficient for experimental fits, but fails to model specific solvent shell changes, such as intercalation.

The first Direct Dynamics project of this work focuses on a bi-metallic Ir complex, where the excited state bond formation results in a large Ir–Ir contraction with oscillatory behaviour. Forty simulated excited state trajectories of 3.5 ps each compare well with experimental results, and uncover a new vibrational mode. We observe how the wide distribution of ground state geometries is responsible for decoherence, and that the solvent cage actually facilitates coherent motion, by blocking the newly discovered vibrational mode. We furthermore observe a non-specific, rotational solvent response to the excitation.

The second Direct Dynamics project studies the effect of solvation on a bi-centred Ru-Co complex, and we observe how the intercalation solvation response which was lost in the study using only force-fields, is recovered in the Direct Dynamics description.

Acknowledgments

As is customary, I would like to start out by thanking my supervisors. However, my appreciation for their supervision, insights, support, ambition, and trust is far from formality. In my days as a Master's student at the University of Copenhagen, Klaus B. Møller's reputation with the students as 'that scary censor from DTU who shouts a lot' preceded him, but I am glad that I still had the courage to convince him to take me on, and now I owe a major part of my academic (and probably also some of my personal) development to his influence.

To Elvar Ö. Jónnson—*Þakka þér fyrir* for all the good times of debugging, help, more help, and more good times. I wonder what this project would look like if we never met, and I would have used another code. I hope for more work (and play) in the future!

In my own department, the help of Niels J. Christensen with the purely classical dynamics simulations has been invaluable, and I hope to repay you soon in GPAW, *and* I hope there will be more time for drum machines and synths in the future!—Still in my department, I would like to extend my thanks to my Ph.D. student 'predecessors', Thomas Kuhlmann and Jacob Petersen always happy to help and hang out. Thanks to Sumera for MD-work.

A big acknowledgement to the experimental part of our group is also more than appropriate. I am grateful for Kasper S. Kjær's help with understanding the ruthenium–cobalt complex. Elisa Biasin's hard work on keeping her Italian temper in check when we did not agree is much appreciated. I fully realize how hard that can be, when—on top of the genetic preordination—you are right in most of the situations, and know it. In any case, if you are going on beamtimes with this crew, which also includes my wonderful childhood friend Tobias Harlang, and Master-of-all-things-Matlab, Tim Brandt van Driel, you are guaranteed to have a great experience. Also, thanks to Kristoffer Haldrup, Morten Christensen, and Martin M. Nielsen for enduring the beamtime spectacle, being there to help and with insights. Morten, thanks for also providing an inspiring soundtrack for the work. The inspiration has carried over into some of the nomenclature, as you may or may not notice.

I have also had a bit of help on the sideline with Python and Fortran questions, since I was still wet behind my coding ears when this project started, so thank you very much Casper Steinmann, Simon Olsson, and 'Ænders' Jacobsen. I hope you won't be afraid to turn on gchat anymore… Thanks also to my native-speaker-consultants Naoise MacNamara and Emma Blake, the latter with whom I've also enjoyed quite a few days at the library, reminding me, when it was most needed, that there is a life outside transient dynamics.

I would also like to thank Karsten Wedel for introducing Elvar and me, and encouraging me to try out GPAW in general. Thanks also to Stephan Sauer for the collaboration on the relativistic effects study, to our experimental collaborators from abroad, and to the IT section back in DTU Chemistry, you guys are extremely flexible and lovely to work with. That's really important.

Thanks to my family for putting up with me during the stressful times: To my father Anders, for lodging and love when I suddenly was without a place to live, and to my mother Hanne. I am pretty impressed with the amount of restraint you have shown toward bugging your only child with all that care and love of yours, during the last, busy weeks!

Finally, thank you for… pretty much everything, Tine Valentin. This will sound incredibly cheesy, but you are the normalization constant to my probability density function: You ensure that, over all space, whatever the time, together we are 1.

Contents

Part I Introduction and Background

1 Introduction 3
1.1 Structural Dynamics and Solvent Responses 4
1.2 Outline 7
References 7

2 The Systems of This Project 9
2.1 Tetracyanoplatinate, $[Pt(CN)_4]^{2-}$ 9
2.2 $M_2(dimen)_4^{2+}$ 10
2.3 $[Fe(bpy)_3]^{2+}$ 14
2.4 $[(bpy)_2Ru^{II}(tpphz)Co^{III}(bpy)_2]^{5+}$: Ru=Co 15
2.5 Summarising the Systems 17
References 18

Part II Preliminary Studies

3 Treating Relativistic Effects in Transition Metal Complexes 23
3.1 The Dirac Equation, and the Pauli- and ZORA Hamiltonians . . . 23
3.2 Effective Core Potentials/Pseudopotentials and PAW 27
3.3 Density Functional Theory 27
3.4 Benchmarking the Methods on Tetracyanoplatinate 30
3.5 Initial Calculations 31
3.6 Comparing the Methods 31
3.7 The Origin of the Relativistic Pt–C Contraction 32
3.8 GPAW Results 33
3.9 Comparison to Experiment 34
3.10 Conclusions 34
References 35

4 **X-Ray Scattering from Purely Classical MD** 37
4.1 Probing Structural Change Using X-Ray Scattering 37
4.2 Molecular Dynamics 38
4.3 X-Ray Scattering from MD Simulations of $[Fe(bpy)_3]^{2+}$ 40
4.3.1 Improvements on the Calculation 42
4.4 An Alternative to the Debye Formulation 43
4.4.1 Testing the Implementation 49
4.5 Returning to $[Fe(bpy)_3]^{2+}$ and Concluding 50
References 53

Part III Direct Dynamics

5 **Background** 57
5.1 The Grid-Based Projector Augmented Wave Method 57
5.1.1 PAW 57
5.1.2 Finite Difference and Real Space Grids 59
5.2 The Combined Quantum Mechanical/Molecular Mechanical Method 60
5.3 Born-Oppenheimer Molecular Dynamics 63
5.3.1 Molecular Dynamics in ASE 64
5.4 Making a QM/MM MD Simulation 66
5.4.1 Preparing Solvent Boxes 66
5.4.2 Analysing Optimal QM Parameters 67
5.4.3 Embedding QM Subsystem in the MM Box 67
5.4.4 The Production Run 67
5.5 Implementing an Acetonitrile Force Field 68
5.5.1 Benchmarking the Acetonitrile Liquid Pseudo-Structure 68
5.5.2 Acetonitrile Dimer Binding Curves 70
References 73

6 **Direct Dynamics Simulations of $Ir_2(dimen)_4^{2+}$** 75
6.1 Preparatory Tests 76
6.1.1 Cell Size 76
6.1.2 Basis Sets and Grid Spacing 76
6.2 Setting up QM $[Ir_2(dimen)_4]^{2+}$ in MM Acetonitrile 79
6.3 $[Ir_2(dimen)_4]^{2+}$ in the Ground State 79
6.4 The Excited State Population of $[Ir_2(dimen)_4]^{2+}$ 81
6.4.1 The Pinching Mode 83
6.4.2 The Twisting Mode 84
6.4.3 The Breathing Mode 84

6.5 Coherence Decay 84
6.5.1 Approximating the Experimental Excitation Bandwidth. 85
6.5.2 The Effect of the Bandwidth on Coherence Time 86
6.6 Coherence Decay and Solvent Interactions. 87
6.6.1 Comparison to Other Systems 89
6.7 The Solvent Shell Response. 90
6.8 Comparing Isomers. 95
6.9 Conclusions. 96
References 97

7 Direct Dynamics Simulations of the Ru=Co Complex 99
7.1 Preparatory Tests 99
7.1.1 Basis Sets and Grid Spacing 99
7.1.2 Vacuum Structural Parameters for the Relevant Electronic States. 101
7.2 Making the QM/MM System 104
7.3 Thermal Distributions 107
7.3.1 Bond Lengths 107
7.3.2 Bite Angles 108
7.3.3 Butterfly Angles. 110
7.3.4 Torsions 111
7.4 Ru=Co Orbitals 111
7.5 Solvent Interaction 113
7.6 Further Discussion and Conclusions 115
References 116

Part IV Summary

8 Summarising Discussion and Outlook 119
References 122

Appendix A: Ru=Co Gas Phase Geometry Optimization 123

Appendix B: Scripts of Relevancy to Future Users 127

List of Figures

Figure 1.1 A Franck-Condon diagram illustrating the general processes of an electronic excitation. The 'S' labels represent 'singlet' surfaces, where there is no unpaired electrons, so the multiplicity is 1 5

Figure 1.2 Illustration of solvation dynamics following electronic excitation of a molecule, which changes its dipole moment. Of course, the solvent arrangement will also respond to structural changes in the solute 6

Figure 2.1 Illustration of the Tetracyanoplatinate complex $[Pt(CN)_4]^{2-}$. The Pt atom is centered between the 4 cyano groups in D_{4h} symmetry, so the complex is *square* planar. . . 10

Figure 2.2 Ligands for the binuclear systems. The complete complexes are made from 2 metal ions, attached at the triple bonded carbon to four opposing ligand units (see Fig. 2.3). Here, the ligands are displayed with increasing flexibility from *left* to *right* 10

Figure 2.3 Schematic of the dimen orientation in the four M_2 $(dimen)_4^{2+}$ isomers. The *highlighted* part of the 2:2 Trans isomer is represented in the schematic in Fig. 2.4. The statistical distribution in solution [11] is displayed below each isomer. Adapted from [12]. 11

Figure 2.4 *Left* The 'windmill-like' structure of the $[M_2(dimen)_4]^{2+}$ complex, here shown without hydrogens for clarity. *Center* Schematic representation which only needs to depict half of the complex, due to its symmetry (see Fig. 2.3). The schematic emphasises the atoms of the 'ligand legs' connected to the metals. This representation will be used in the explanation of the simulated dynamics

in Chap. 6. *Right* Molecular orbital diagram, reconstructed from [15], featuring the metal-centered, bonding LUMO and the antibonding HOMO 11
Figure 2.5 *Left* Deformational isomerism in $[Ir_2(dimen)_4]^{2+}$. The 'Short and Twisted' conformer can only be theoretically described by mapping energy surfaces of the Ir–Ir distance and the twisting angle between the metal ligand planes, while allowing the rest of the structure to relax at each point. *Right* Absorption spectrum of $[Ir_2(dimen)_4]^{2+}$ in acetonitrile: Pumping the molecule at 475 or 585 nm selectively excites the long and eclipsed, or the short and twisted conformer, respectively [19]. 12
Figure 2.6 Ground- and excited-state energies (*left* and *right*, respectively) of the 3:1 isomer of $[Ir_2(dimen)_4]^{2+}$, evaluated in vacuum as a function of Ir–Ir distance and dihedral angle of the opposing ligands. While the GS potential is very shallow, there is an indication of a minimum around 3 Å and 35°. Reproduced with permission from van Driel and Nielsen [17] 13
Figure 2.7 Ultrafast Time-resolved Spectroscopy studies of $[Ir_2(dimen)_4]^{2+}$ in acetonitrile solution. *Left* Transient absorption, pumped at 527 nm [12]. *Middle* Stimulated Emission from pumping at 590 (short and twisted) [19]. *Right* Stimulated Emission pumped at 477 nm (long–eclipsed). Restating the article, the electronic decay of the signal was fitted to a multiexponential decay. The residual between this decay and the signal was Fourier transformed for time delays after $t \geq 0.5$ ps to get a vibrational frequency of 75 cm^{-1}. 13
Figure 2.8 *Left* A Ball-and-stick model of the $[Fe(bpy)_3]^{2+}$complex, taken from Daku et al., using their definitions of angles important for the overall structure of the complex [24]. *Right* Schematic drawing of the states believed to take part in the spin crossover dynamics, as represented in [25], where changes in spin induces changes in Fe-ligand bond length. 14
Figure 2.9 The basic principle of a Dye-Sensitized Solar Cell (DSSC). The semiconductor material is photosensitized by a dye adsorbed to the surface, which, when photoexcited, injects an electron into it, into the anode. This oxidizes the dye, which then in turn oxidizes the electrolyte mediator, which is then reduced by the cathode. 15

Figure 2.10 The $[(bpy)_2Ru^{II}(tpphz)Co^{III}(bpy)_2]^{5+}$complex, where (tpphz) = tetrapyrido (3,2-a:2′3′-c:3″,2″-h::2‴,3‴ $-j$) phenazine [39]. For brevity, the molecule is called Ru=Co throughout this work 16
Figure 2.11 The various electronic valence configurations of Co-systems for their common multiplicities and oxidation states 16
Figure 3.1 Scheme illustrating the differences of the spectra of the non-relativistic and relativistic Hamiltonian. By solving the Dirac equation, apart from a shift in zero-point for the energy, one also obtains a negative continuum of unbound states 25
Figure 3.2 Benchmark of relativistic approximations on the Tetracyanoplatinate geometry. The scale of the Pt–C plot is ten times larger than the C–N plot. Neglecting any type of relativistic treatment has a large impact on the Pt–C bond. The labels 'Dyall.v3z', 'aug-cc-pVTZ', and 'def2-TZVP' do all correspond to various triple-zeta basis sets [27–33], and their corresponding ECP [34, 35]. 32
Figure 3.3 $[Pt(CN)_4]^{2-}$ Bond lengths as a function of real-space grid spacing h, and basis set size. *Left* Pt–C, *Right* C–N. The bold line on both plots represent the value $h = 0.18$ Å, which is often employed in GPAW 33
Figure 4.1 An Illustration of the parametrised energy terms of Eq. 4.3. The indices a, b, t, m, and n run over the angles, bonds, torsions and atoms, respectively, and 'eq' represents the equilibrium value. 39
Figure 4.2 Test of the modified equilibration scheme of Desmond, which keeps the solute frozen under all equilibration steps (the *red*, *blue*, and *green curves* on the *left* grafs), and also in the production run. The restraining is done by adding another potential $E_{res} = K_{res}(r - r_{res})^2$ to the total energy. For the equilibration, $K = 1000$ kcal/(mol · Å^2), and for the production, $K = 500$ kcal/(mol · Å^2) 40
Figure 4.3 Histograms of each of the 6 Fe–N bonds from the MD simulations using the 500 kcal/mol restraint on all $[Fe(bpy)_3]^{2+}$distances. *Blue* The GS geometry. *Red* ES geometry. 41
Figure 4.4 Plots of the calculated scattering from snapshots of the MD trajectories. The *black curves* represent the averaged signal. Spurious oscillations with high-frequencies are observed at low q, originating from the unphysical truncation of the simulation box 42

Figure 4.5 *Left* The total XDS signal at $t = 1$ ps, with the fit carried out by the experimental part of our group [I]. *Right* The 'Solute+Cage' signal, i.e. the calculated average difference signal from the two trajectories containing ensembles of solvent configurations for each of the frozen structures (*green curve*). The *dashed grey curve* is calculated difference scattering from the naked solute structures obtained from [8], showing a dramatic change in the signal if the solvent is neglected 42

Figure 4.6 Comparison of numerical implementations of calculated X-ray scattering, $S(q)$, implemented in Matlab. The scattered intensity goes to the number of electrons in the sample squared, when $q \rightarrow 0$. The estimated standard error of the calculation using the $g(r)$ with $\Delta r = 0.001$ Å is $2.12 \cdot 10^{-6}$. The *green curve shows* the residual. The *blue dashed curve* shows the residual when using $\Delta r = 0.1$ Å, giving a standard error of $1.37 \cdot 10^{-4}$. The *inset* shows the $g(r_{\Delta=0.001})$-function used in the calculation 49

Figure 4.7 *Left* Scattered X-ray intensity from neat water, experimentally obtained by Hura et al. (*blue line*) [43, 44]. From that experiment Hura et al. resolved the O–O pairwise RDF, which was then used in our numerical implementation, to give the *dashed black curve*. The *dashed green line* represents scattering calculated via pairwise RDF_s obtained by Mahoney and Jorgensen [46]. This data lacks the delta-functions from intramolecular O–H and H–H correlations, which causes the errors in low-q. *Right* A figure from Hura et al. [44], comparing calculated scattering within the IAM (*black line*) and with a Modified Atomic Form Factor, that takes chemical bonding into account 50

Figure 4.8 *Left* Difference scattering signal calculated from MD simulations redone in the NVT ensemble, compared to Debye-scattering from the first MD runs. *Right* The full signals 51

Figure 4.9 *Left* the RDF_s obtained via an ab initio MD method [8]. *Right* RDF_s from the MM MD simulations based on the molecular structures obtained from [8]. While the two models produce qualitatively similar results, the specific solvation dynamics are not reproduced, since there is effectively no change in coordination number in the first solvation shell, in the MM model 52

Figure 5.1 The basics of GPAW. The pseudo-wave function is described numerically on an uniform grid, and the all electron core states within the augmentation region are described on a finer, radial grid, and can be evaluated independently of the neighbouring environment, before the actual calculation, and saved in a GPAW 'setup'. The PAW method makes sure that the two regions are smoothly matched 58
Figure 5.2 An example of how the 6s orbital of Pt is computed within the PAW formulation. The plot shows the real values of the various wave functions as a function of the distance x from origo. The a superscript demarcates the wave functions within the atom-specific 'augmentation sphere', and the tilde-marked are the pseudo-wave functions. From [7]. 58
Figure 5.3 A one-dimensional grid used as an example for the 3D mesh used in GPAW 59
Figure 5.4 A Schematic representation of the main idea behind the hybrid QM/MM strategy. The total system to be simulated is divided up in two subsystems: *1* The electrostatically embedded QM subsystem, where the electronic structure is calculated using DFT. *2* The MM subsystem which is described using classical force-field methods. The two subsystems are interacting according to the description in the text. 61
Figure 5.5 *Left* Schematic of the periodic boundary conditions. *Right* Schematic of how long-range electrostatics are cut off in the MIC. The highlighted molecule is electrostatically affected by the fully drawn molecules, comprising a minimum image with the original box dimensions. 62
Figure 5.6 Plots of the effect of increasing the friction term γ in Eq. 5.16, illustrated by a simple numerical implementation of the equation, using a harmonic potential. The same integrators as in ASE is used. $\gamma = [0; 0.001; 0.01; 0.1; 0.5]$, from *top* to *bottom*. 64
Figure 5.7 Schematic of the QM/MM code; how it operates within ASE, and together with GPAW. The main interfacing takes place in ase_qmmm, where the two ASE objects are created from the subsystems, the PBC are handled in the presence of the QM cell, and calls to the calculators are defined. calc_mm is the calculator for the MM subsystem, and calc_ks for the QM subsystem. So far, the interfacing

works together with the ASE dynamics module, but could be readily expanded to work with other tasks, such as geometry relaxation, ΔSCF methods for approximation of excited states [22, 23], etc. Adapted from [4] 65
Figure 5.8 *Left* Histogram of all the forces in the QM/MM simulations of a single QM ACN in an MM ACN box. The QM-to-MM interfacing is working properly if there is no average force on the entire system. The avg. force from the simulation is 4.41×10^{-8} eV/Å. *Right* A histogram of the translational speeds of the center of mass of each ACN molecule in the purely classical NVT simulation, plotted with the theoretical Maxwell-Boltzmann distribution 69
Figure 5.9 Radial distribution functions (RDF) of the 6 intermolecular distances of ACN. The RDF are calculated every 500th fs in trajectories of 0.5 and 0.25 ns for the MM and QM/MM systems, respectively. The QM/MM systems are comprised of a single QM ACN in a bath of MM ACN. The QM/MM1 systems are made using methyl-group vdW parameters for the middle carbon [24], while QM/MM2 uses aliphatic parameters [25]. There is overall a good agreement between the various descriptions. 70
Figure 5.10 Lennard-Jones potential for the nonbonded Me-N interactions in acetonitrile, combined using the Lorentz-Berthelot- and Waldman-Hagler rules, respectively. The difference is largest in the very repulsive region 71
Figure 5.11 Two known ACN dimers, where the cyclic version is believed to have the lowest energy [30–33]. The dimer binding energy is obtained by subtracting 2 times the potential energy of the single molecule from the total potential energy of the dimer, at increasing distances *d*. 71
Figure 5.12 Binding energy curves of ACN. *Top* The cyclic dimer. *Bottom* The linear dimer. Since there are two ways of combining the QM and the MM description, both the 'linear' and the 'linear reverse' dimers have been defined as in the graph, and examined. The insets depict are zoom-ins on the largest differences between the MM/MM and QM/MM curves. The 'FD' curves are made in the pure grid-based finite difference mode, whereas the LCAO-labeled curves uses basis sets from linear combinations of atomic orbitals for the initial wave function guesses 72

Figure 6.1 *Left* Two of the main dynamic, structural modes of $[Ir_2(dimen)_4]^{2+}$, here shown without hydrogens for clarity, for the long and eclipsed conformer, as described in Chap. 2, p. 10 and on. *Right* Sketch of a Franck-Condon diagram, showing how experimental excitation is approximated in the simulations 76

Figure 6.2 Ir–Ir distance as a function of cell size. The relaxation does not converge for cells with 3 Å of vacuum, but for vacuum padding of 4 Å or more, the Ir–Ir changes are within the accuracy of the real space method. The calculation with the largest cell did not converge within the allowed walltime. The calculations were carried out using a grid spacing of 0.18 Å and a tzp/dzp basis set for Ir/the rest of the molecule 77

Figure 6.3 Convergence of the main structural parameter of the complex—the Ir–Ir distance—with respect to the real space grid spacing h, and size of basis set. A stable value for the distance is maintained when employing values of $h < 0.22$ Å with basis sets of at least double-zeta size. The convergence criteria for each individual relaxation was a maximum force of 0.05 eV/Å on any of the atoms in the system 77

Figure 6.4 Convergence test of the Ir–Ir distance using the optPBE-vdW functional. While the fluctuations in distance are an order of magnitude smaller than with the normal PBE, the convergence is not as convincing as with PBE. Furthermore, the mean distance in the most stable southeastern quadrant is still overshooting the experimental value. At last, this functional had many convergence problems, as can be seen from the plot of the average forces in the system, where only a few of the calculations reached forces smaller than the convergence criterium 78

Figure 6.5 *Left* Re-equilibrating the system after adding $[Ir_2(dimen)_4]^{2+}$. In first ~2.5 ps, the complex was also coupled to the Langevin thermostat. The part of the trajectory after the black stipled line was used as the production run for the ES simulations. *Right* The center-of-mass speed distribution of all ACN_s, sampled from the start of the production run. The avg. forces on the entire system is 2.36 μeV/Å per atom. 79

Figure 6.6 Structural modes in $[Ir_2(dimen)_4]^{2+}$ plotted for the GS trajectory. *Top* Ir–Ir distance. *Middle* Mean dihedral angle. *Bottom* Mean distance between opposite carbon atoms in the ligand rings. This parameter describes a breathing motion which will play an important role in the excited state (see Sect. 6.4.3 and on). The stipled lines indicate the allowed region within k_BT of the energy surfaces from [14] (see Fig. 2.6). The ES runs were started from the configurations marked with black crosses 80

Figure 6.7 Analysis of energetic drift, defined as the total energy $E_{tot}(t)$ minus the total energy at $t = 0$ after excitation, E_{tot}. The main energy increase is apparent in both un- and -thermostated simulations, which means that the observed increase in energy is due to the excitation 81

Figure 6.8 Visualization of the simulation. The *blue* and *red* isosurfaces depict the HOMO in each case, and the grey isosurface is the total electronic density of the QM subsystem. All isovalues are identical between the two snapshots. *1* The complex in the ground state. The GS HOMO does not have any spacial overlap between the metal atoms. *2* A snapshot of one of the ES trajectories, at the point in time where the Ir–Ir contraction is largest. Here, we observe overlap of both the HOMO (the GS LUMO) and the resulting total electronic density 81

Figure 6.9 The Ir–Ir pinching (*top*), twisting (*middle*), and breathing (*bottom*) as a function of time after excitation. The twist and the breathing is defined as the change in mean dihedral angle and in mean distance between opposite ring-carbons in the ligands, respectively, from the moment of excitation. The *black curves* are the averages, while the shaded backgrounds show the binned counts of values from all trajectories. The bin sizes are 50 fs by 0.1 Å, 0.05 Å, and 2 degrees for the pinching, breathing and twisting, respectively. This population loses its pinch phase within a picosecond, while the single trajectory-pinch oscillation dies out much slower, as can be seen on the top inset. However, the mean phase of the breathing mode is sustained for almost 1.5 ps. The pinching- and breathing modes share the same oscillation period 82

Figure 6.10 The change in distance of the atoms oppposite to each other, adjacent to Ir. The two atoms in each ligand closest to Ir are also pinching, while the next carbon (C_1) is almost unaffected 83

Figure 6.11 Approximation of the effective bandwidth used in creating the excited state population. The excitation energy is approximated by subtracting the single point energy of each initial GS configuration from a single point calculation of its ES spin configuration counterpart.

The *lower inset* shows histogram of these energies, converted to wavelengths, and how many times they appear in the ES population. This is extrapolated to the entire GS trajectory, under the (crude) assumption that the excitation energy is linearly dependent on (and only on) the Ir–Ir distance, shown in the *top inset*. This relation is then used for obtaining the results shown in the main figure 85

Figure 6.12 Splitting up the total ES population with respect to excitation energies, or by grouping initial configurations. The *top* two ES pinch graphs are made from selecting for ES trajectories with excitation energies assumed close to the experimental value and bandwidth. The *bottom left* subset is made from a narrow internal phase space of initial Ir–Ir distances. The *last graph* shows the dynamics of a wide phase space, but with the same statistics as the other three subsets 87

Figure 6.13 *Left* Dynamics of the complex of progressively more gas-phase character (from *bottom* to *top*). *1* Pinch of the fully solvated complex. *2* A simulation where the solvent was removed at the time of excitation, but keeping the initial geometry influenced by the solvent, *and* the velocity of the atoms in the complex. *3* A simulation where only the geometry is kept, while the initial velocities are set to zero. *4* An excitation started from the gas phase, zero kelvin structure of the complex. *Right* Comparison of the average distance of the opposing ligands carbons that define the breathing mode. The average solvated pinch (*dashed blue curve*) is overlaid for timing comparisons. The *dashed black line* is the mean GS ring carbon distance 88

Figure 6.14 Comparison of the breathing mode in the gas phase trajectory (*red*) and two trajectories of solvated systems, displaying the energy transfer mechanisms in the system. The two solvent-included trajectories are started from similar GS conditions, but 10 ps apart, so the two accompanying solvent configurations are uncorrelated. The breathing mode amplitude is large in gas phase, compared to the other trajectories, since there is no dampening from the solvent. The small pertubations in the breathing oscillation is attributed to solute-solvent interactions, governing the vibrational relaxation through the solvent 89

Figure 6.15 Solvent-related $\mathrm{RDF_s}$. *Left side* Between the iridium atoms and each end atom of the solvent molecules. *Right side* Between the dimen ligand carbons and each end of the solvent molecules. The RDF_s are sampled with $\delta r = 0.2$ Å at each time step over all the ES trajectories, and temporally averaged with a bin size of 50 fs, and finally smoothed with a 5 point moving average 90
Figure 6.16 The development of averaged RDF values over time. The time-binning has a width of 100 fs, and the r span is: *Top* Ir–Me r from 9 to 10.2 Å. *Middle* Ir–N from 8 to 9 Å. *Bottom* Ir–N from 4 to 8 Å. 92
Figure 6.17 The evolution of Ir-Solvent coordination numbers with time. The marked areas of the $\mathrm{RDF_s}$ in the middle graphs demarcate the regions used for the the plots of the temporal evolution of the short-r and long-r region (*left* and *right* plots), respectively. The *insets* show the running coordination numbers, $c_n(r)$ and the *dashed line* shows the end value chosen for the short-r region. The $c_n(r)$ slope increases around 6.1 Å for $c_{n,\mathrm{Ir-Me}}(r)$, indicating that the $g(r)$ is now also sampling coordinating solvent molecules on the other side of the complex. For the less coordinated nitrogen ends, the slope change is more subtle, but estimated to 6.5 Å. The right region starts from the end of the left, and is chosen to end when all the $c_n(r)$s are as similar as possible, indicating the bulk region has been reached. The time-bins for averaging is 20 fs 92
Figure 6.18 Coordination number ratios of Ir–N:Ir–Me (*top*), and Ligand-N:Ligand-Me (*bottom*), averaged in 400 fs bins. For iridium, the GS ratio is below 1 for short distances, but this ratio significantly increases after the first 400 fs. For the ligands, no significant changes are seen, and the ratio always above 1. In the long-distance limit of the bulk solvent, the ratios converge to 1, which corresponds to the random orientation. The *insets* in the *bottom plot* depict snapshots of all the full ACN molecules within 4 Å of the solute 93
Figure 6.19 2D plot of the temporal evolution of the $\mathrm{RDF_s}$ between Ir and each end of ACN. In the *top plot*, r for the peak increases with time, in conjunction with a decreasing $g(r)$-value. In the *bottom plot* the grow-in of a peak at short distances occur after roughly 2 ps 94

Figure 6.20 The four $M_2(dimen)_4^{2+}$ Isomers as described in Sect. 2.2. The All simulations up until now has been carried out on the 2:2 Trans Isomer 95
Figure 6.21 Gas phase dynamics of all four $[Ir_2(dimen)_4]^{2+}$ isomers. All three main modes are expressed in all isomers, and the differences are at most on the same scale as the changes invoked by the different solvent configurations of the solvated simulations 95
Figure 7.1 *Left* Definitions of the relevant geometric parameters for the Ru=Co complex. The Ru-centre is to the *left*, and the Co-centre to the *right* side of the image. 'Me' (Metal) is either Ru or Co. While the bite angles are only depicted for the Co-centre, they are of course defined likewise for the Ru-centre, and vice versa for the torsions, bond lengths and butterfly angle. *Right* Labeling of the three electronic configurations simulated in this work 100
Figure 7.2 Convergence of the Ru–N distances as a function of grid spacing *h* and basis set size. The *color bar* shows the distance in Å. 100
Figure 7.3 Convergence of the Co–N distances as a function of h and basis set size. The *color bar* shows the distance in Å. 101
Figure 7.4 Vacuum relaxed GS Co–N bond lengths (*blue lines*). The *green* graph represents the average force on the molecule, and the *red* is the average forces on the nitrogen atoms. For the GS, Convergence is reached within 6 steps, and the bond length changes are maximum 1 pm 102
Figure 7.5 Vacuum relaxed GS Ru–N bond lengths (*blue lines*), with the *red* and *green curve* displaying the avg. forces on the nitrogens and the entire molecule, respectively 102
Figure 7.6 Vacuum relaxed LS Co–N bond lengths. While there is a constant strain on the nitrogen atoms, the distances seem to converge to constant values. 103
Figure 7.7 Vacuum relaxed LS Ru–N bond lengths, where the same trend as for the Co–N bonds is observed 103
Figure 7.8 Vacuum relaxed HS Co–N bond lengths, showing the same behaviour as the LS state. The Co–N_B bonds seem to have reached a stable distance, as well as the Co–N bonds 3 and 6. 2 and 3 seem to be close to convergence 104
Figure 7.9 Vacuum relaxed HS Ru–N bond lengths. 104
Figure 7.10 The solvent temperature of the Ru=Co complex in the electronic ground state, showing the initial equillibration of the temperature. The first 130 fs are simulated using 0.25 fs steps at 50 K, with a friction of 0.1 on the entire system. The friction is set to 0.05 on the solvent only, for

the production run-part. The *left inset* shows a zoom of the first ps. The *right inset* shows the actual versus theoretical temperature distribution of an NVT ensemble of finite size, as discussed in Sect. 4.3 105
Figure 7.11 Plots of the equillibration of the main structural parameters in the system, the Metal-N distances. The *top* two plots show the 6 Co–N distances for run C and H. The *bottom* two plots are for Ru–N. The distances have been smoothed with a 0.5 ps moving average to reveal any possible change on longer time scales. The Co–N dynamics in the first ps are shown in the inset in the *top* two plots, to more clearly reveal the initial equilibration process. 106
Figure 7.12 50 % Probability- or thermal ellipsoid representation of the GS average structure of Ru=Co, made from run H. The representation is made by aligning the Ru=Co geometry to the first frame, to remove rotations and translations of the entire molecule, using the RMSD Trajectory tool in VMD [5], and then exported to a new trajectory of aligned frames, from where the covariance matrix was calculated, exported in the 'anisou'-fields in a pdb file, and visualised using Chimera [6] 107
Figure 7.13 Evolution of the Metal-N bonds in Ru=Co. The data has been re-binned in 10 fs averages, and each distance after the first is raised with 0.1 Å more than the previous distance, so that the individual trajectories can be seen. No vibrational cooling is observed for the LS excitation, but for the HS, the vibrational motion in both metal centres is seen to cool down in roughly 6 ps 108
Figure 7.14 Thermal Distributions of the Ru–N (*top row*) and Co–N (*bottom row*) bonds. The x-axis marks represent the vacuum values in the same colour code as the thermal distributions. ΔR is the average bond length change of each Co–N bond, w.r.t the average GS bond length of each bond. In the Ru-centre, almost no change in average bond length from GS to the excited states are seen, the maximum being a 0.03 Å contraction in the HS state, for two of the bonds. The gas-phase lengths are very similar to the solvated averages. The six HS distributions are much narrower, with an average standard deviation of 0.02 Å, compared to 0.09 Å and 0.06 Å for GS and LS, respectively. For the Co–N bonds, an average lengthening of 0.14 Å is observed in the HS state. In the LS state, some

bond elongation seems to occur, and the thermal distribution becomes even wider, with indications of double peaks or shoulders. Finally, the LS solvated average bond lengths are longer than the vacuum values 109
Figure 7.15 LS Thermal distributions of a Co–N_{B1} bond (*left*), a nd Co–N_3 (*right*). The distribution of the non-bridge bonds split up, where one part has bond lengths more comparable to the HS state 110
Figure 7.16 The Ru=Co bite angles of the Co centre, sampled in bins of 0.5°. The HS distribution becomes narrower, similar to the bond length distribution, and all three bite-angles decrease $\sim 5°$ in the HS state. The Ru–N centre results are not shown, since the angles remain largely unchanged. The *dashed red lines* show the gas phase results, while the *dashed white lines* show the average of the solvated distribution 110
Figure 7.17 The butterfly angles θ of both the Ru- and Co-centre. The angles remain largely unchanged by both excitation and/or solvation. 110
Figure 7.18 The torsion angles for the Co-centre (*top*) and the Ru-centre (*bottom*), binned in 0.5° bins. The excited state angles for the Co-centre are slightly larger than for the ground state. For the HS Ru-centre, the solvation seems to bring the angle back to the GS values . . . 111
Figure 7.19 Overview of the relevent occupied and virtual orbitals for Ru=Co, in the ground state, and the two ES spin manifolds of LS and HS. We observe that the GS LUMO is similar to the LS and HS α spin HOMO. The energy levels are sampled from each time step, and are seen to be distributed with a variance of 10 k_bT. Nothing has been done to improve on the absolute energies of the virtual orbitals, and as such they are mostly displayed for completeness. The ES β-spin channels suffer from substantial charge-smearing, as shown in the lower left inset. This is due to the finite Fermi-Dirac width which had to be applied for converging the SCF-optimizations within a feasible amount of steps for the QM/MM MD simulations 112
Figure 7.20 Pairwise RDF between each of the metals and each end of the solvent molecules. The *dashed lines* indicate the chosen solvent shell definitions. The RDF are sampled from the equilibrated parts of the three trajectories, with a

Δr of 0.2 Å and are shown here after a 3 point moving average smoothing. The N-values are coordination numbers of each state, and ΔN are the differences from the GS to the two excited states. The Ru-centre increases its coordination number in the HS state. 113
Figure 7.21 Cumulative coordination numbers, $c_n(r)$ of the Metal-Solvent correlations 114
Figure 7.22 The N:CH_3 ratio of the cumulative coordination numbers for Co (*top*), and Ru (*bottom*). The two images depict snapshots of the GS (top) and HS (*bottom*) solvent shells, defined by drawing only the full solvent molecules which are within 8 Å of either Ru (*left on the images*) or Co. The ratios converge to 1 at long distances, as expected. 115
Figure 8.1 The first step of implementing a polarisable force-field in ASE. *Left* O–O Dimer PES, from the ASE-interfacing of the original force field code. *Right* Results from the original force-field paper [1] 120

List of Tables

Table 3.1	Deviations from the 4-component results for the optimized Pt–C and C–N bond lengths of $[Pt(CN)_4]^{2-}$ in pm	34
Table 5.1	The non-bonded Molecular Mechanics parameters, and intermolecular fixed distances, from Guardia et al. [24] ϵ is the potential well-depth and σ the interaction distance. . . .	68
Table 7.1	Collected structural parameters for Ru=Co in gas phase.	105

Abbreviations

ACN	Acetonitrile
ASE	Atomic Simulation Environment
BSSE	Basis Set Superposition Error
DFT	Density Functional Theory
DSSC	Dye-Sensitized Solar Cell
ECP	Effective Core Potential
ES	Excited State
ET	Electron Transfer
EVR	External Vibrational Relaxation
FD	Finite Difference
GGA	Generalized Gradient Approximation
GPAW	Grid-Based Projector Augmented Wave method
GS	Ground State
HF	Hartree-Fock
HOMO	Highest Occupied Molecular Orbital
HS	High Spin
IAM	Independent Atom Model
IC	Internal Conversion
ISC	Intersystem Crossing
IVR	Intramolecular Vibrational energy Redistribution
KS	Kohn-Sham
LCAO	Linear Combination of Atomic Orbitals
LCLS	Linac Coherent Light Source
LJ	Lennard-Jones
LS	Low Spin
LUMO	Lowest Unoccupied Molecular Orbital
MAFF	Modified Atomic Form Factor
MD	Molecular Dynamics
MIC	Minimum Image Convention
MLCT	Metal-to-Ligand Charge Transfer
MM	Molecular Mechanical

PAW	Projector Augmented Wave method
PBC	Periodic Boundary Condition
PBE	Perdew-Burke-Ernzerhof
PES	Potential Energy Surface
QM	Quantum Mechanical
QM/MM	Quantum Mechanical/Molecular Mechanical
RDF	Radial Distribution Function
SCF	Self-Consistent Field
TDSE	Time-Dependent Schrödinger Equation
vdW	van der Waals
VR	Vibrational Relaxation
XDS	X-ray Diffuse Scattering
XFEL	X-ray Free Electron Laser
ZORA	Zeroth Order Regular Approximation

Part I
Introduction and Background

Chapter 1
Introduction

The work presented in this thesis is focused on describing how the dynamics behind nuclear motion play out on the stage of femtochemistry. The femtosecond timescale is the fundamental timescale on which chemical reactions occur, since it is the timescale of nuclear vibrational motion. The pioneering experimental work on uncovering the processes of the chemical reaction in its own temporal domain was carried out by Ahmed Zewail and co-workers, for which he received the Nobel prize in 1999 [1], underscoring the importance of the field. The seminal work used transient spectroscopy to follow the dissociation of diatomic [2] and triatomic [3, 4] molecules in real time. The ultrashort light pulses of the spectroscopic *probe* provides the necessary 'shutter speed' to catch the molecular motion, but just as a film director needs a clapperboard to synchronise sound and vision, the motion of millions of billions of molecules must be synchronised. In this way, the averaged behaviour of molecules in the sample retains information of the behaviour of each individual molecule. This is achieved by using laser *pump*-pulses of few-femtosecond durations to initiate the process. The combination of the *pump*- and *probe* tools define the femtochemical *pump-probe* experiment, where the spectroscopic methods can be used to interpret the kinetics and the changes in the energies of electronic states of the system. While they can be informative in their own right, they do not necessarily have a 1:1 correspondence to the changes undergone by the molecular structure, and thus do not automatically shed light on the structure/function relationships in the studied systems. However, less ambiguous information on the (changes in) molecular geometries can be obtained by using X-ray scattering as the probe instead. Recently, the limits of what processes are experimentally accessible to X-ray Diffuse Scattering (XDS) in a pump-probe setup has been pushed back almost 3 orders of magnitude, with the introduction of X-ray Free Electron Lasers (XFELs), large-scale facilities capable of producing <100 fs X-ray pulses of formidable intensity.

Whereas the ordered nature of solid state materials utilize the diffraction condition to precisely reconstruct crystal structures, the interpretation of a time-resolved XDS signal from molecules in solvent introduces further challenges on the interpretation of the signal, that emphasizes the necessity of theoretical modelling. Continuously working towards discovering the nature of the chemical reaction entails scrutinizing

A.O. Dohn, *Transient Changes in Molecular Geometries and How to Model Them*, Springer Theses, DOI 10.1007/978-3-319-18747-1_1

molecular systems of increasing size and complexity, which again motivates the need for development of theoretical tools that feasibly model the nature of the processes expressed in the experimental systems.

The main part of the work presented in this thesis is thus focused on employing, benchmarking, and helping the development of a 'Direct Dynamics'-model that combines the quantum mechanical description of Density Functional Theory (DFT) with the classical Molecular Mechanical method that successfully and computationally feasibly can represent the processes exhibited in molecular complexes of significant size and intricacy. The name 'Direct Dynamics' arises from the notion that the atomic motion is simulated by directly calculating the classical forces on the nuclei, influenced by the explicitly calculated electronic density of the Quantum Mechanical part of the system. This calculation is again affected by the classical point charges of the surrounding MM part. The resulting atomic motion is collected—together with the associated velocities—into *trajectories* which can be analysed to reveal information about the relevant processes undergone by the system.

The following section will give a brief and general overview of these dynamical processes of which some has been the focus of this work.

1.1 Structural Dynamics and Solvent Responses

The equation that fully describes a molecular system is the Time-Dependent Schrödinger Equation (TDSE) [5]:

$$i\frac{\partial}{\partial t}\Psi(\mathbf{r}, t) = \mathbf{H}\Psi(\mathbf{r}, t) \tag{1.1}$$

where Ψ is the total nuclear and electronic wave function of the molecular system, and $\mathbf{H}$ is the Hamilton operator. The general solution of the Analytical solutions to the TDSE is an arbitrary superposition of any number of particular functions that satisfy the equation, called a wave packet, which propagates with time, in some cases coherently, meaning that the spread of the wave packet stays the same [5]. Analytical solutions to the Analytical solutions to the TDSE only exist for the simplest of systems. However, the time-dependence can be separated out, and due to their large difference in mass, the nuclei can be assumed to appear stationary to the electrons, such that the motion of the two types of particles also can be separated [6, 7]. This is called the Born-Oppenheimer approximation, and it means that any small change in the nuclear positions will instantly be mirrored in the electronic structure, and as such, the nuclei move in a potential of the (instantly corresponding) electrons. For a diatomic molecule, this potential will simply be some function of the distance of the two nuclei, but for an arbitrary number of nuclei, this potential is called the Potential Energy Surface (PES), where the surface is N-dimensional, with N being the degrees of freedom of the molecule.

The curves in Fig. 1.1 represent PESs of a molecule, where the multidimensional hypersurface is simplified as a 1D curve. The state labeled S_0 represents the electronic

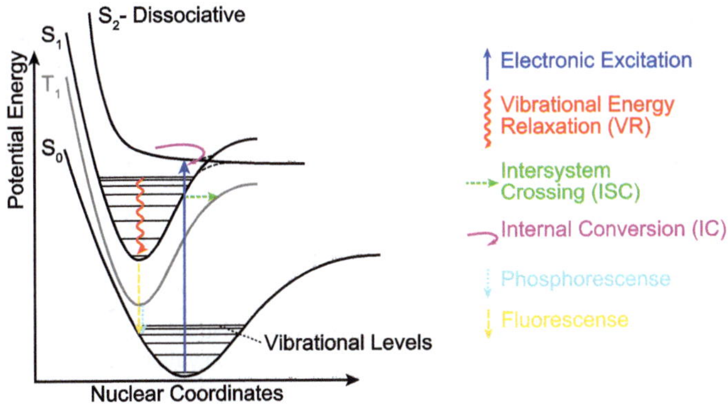

Fig. 1.1 A Franck-Condon diagram illustrating the general processes of an electronic excitation. The 'S' labels represent 'singlet' surfaces, where there is no unpaired electrons, so the multiplicity is 1

ground state with a potential energy minimum for a given set of interatomic distances and angles, or simply 'Nuclear coordinates'. At 0 K the molecule can only exist in the vibrationally lowest level, but at increasing temperatures, more vibrational levels gradually become available. When a molecule is electronically excited, e.g. by absorbing an incoming photon, an electronic transition following the Franck-Condon principle [8] takes place, which is illustrated by the vertical blue arrow. Depending on the photon energy, the resulting electronic transition can be to any of the higher electronic states, in any of their vibrational levels. Some can be associative, as exemplified in the S_1 surface in Fig. 1.1, while others can be dissociative, e.g. like S_2. An actual excitation is often to a higher-lying electronic state, which is quickly relaxed to the S_1. This initial, very fast relaxation cascade involves processes where the nuclei do not appear stationary to the electrons, so that electronic motion cannot be separated, and the description of these Internal Conversion (IC) processes fall beyond the Born-Oppenheimer approximation. Theoretical treatment of such so-called *non-adiabatic* processes are currently limited to systems of much fewer atoms than the ones in this work.

Another possible route in the relaxation cascade is Intersystem Crossing (ISC), where the spin of the electron flips, caused by coupling of the intrinsic spin of the electron to its orbital angular momentum, so that the overall multiplicity of the system, $2S + 1$, where S is the total spin momentum, changes. The grey T_1 is an example of a triplet surface, where there is two unpaired electrons which each has spin $s = \frac{1}{2}$, giving the total spin momentum $S = 1$, so the multiplicity is 3.

In any case, the longest-lived state of the excitation will often be the final excited state [9], within a time scale accessible to experiment. The excess energy of the excitation is dissipated by non-radiative, dynamic processes termed Vibrational Relaxation (VR): Vibrational modes of the complex in which the structure dynamically oscillates according to the potential it experiences. This process can be further divided into two main channels of dissipation:

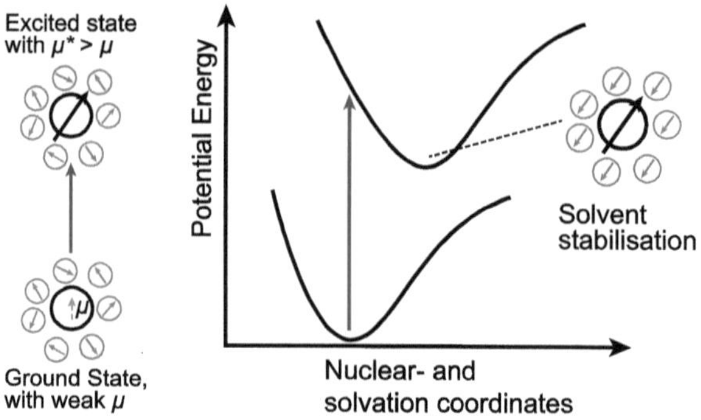

Fig. 1.2 Illustration of solvation dynamics following electronic excitation of a molecule, which changes its dipole moment. Of course, the solvent arrangement will also respond to structural changes in the solute

1. Intramolecular Vibrational energy Redistribution (IVR)
2. External Vibrational Relaxation (EVR)

where IVR (as the name suggests) takes place within the molecule, redistributing the excess energy into different vibrational modes. In EVR, the energy is dissipated to the solvent, e.g. via collisions with solvent molecules.

Since electronic excitation events often lead to significant changes in the electronic structure of the molecule, its overall dipole moment μ will change in those cases. For polar solvents, this means that the solvent molecules closest to the solute will respond to this overall change in electronic structure, which exemplifies apart of the EVR process, here termed the non-specific solvent response.

Figure 1.2 illustrates the corresponding solvent re-orientation dynamics, wherein reciprocally, the solvent can stabilise (i.e. lower the energy of) certain Excited State (ES) nuclear coordinates, such that the solvated structure differs from the gas-phase. Additionally, the figure depicts how the structural *change* started by the electronic excitation can be different in a solvated system, compared to the gas-phase equivalent.

Another option for solvent response go under the label of 'specific solvation dynamics', where the closest solvent molecules which might be intercalated in the solute are sterically forced away from the solvent shell (i.e. the innermost region of solvent molecules), or conversely, changes in the solute geometry can open up for more solvent molecules to intercalate, thus explicitly changing the solute-solvent interaction of a well-defined set of molecules.

As previously mentioned, the choice of theoretical tools for studying these processes will always depend on the size and complexity of the system. The systems chosen for this project are described in the following section. They all exhibit (a combination of) the just described processes. Therefore, examining the systems can provide answers to the fundamental questions: *How is excess energy from electronic excitations dissipated? What is the role of the solvent?* The examination in this work consists of Direct Dynamics simulations, also termed Quantum Mechanical/Molecular Mechanical (QM/MM) Molecular Dynamics (MD) simulations, as previously mentioned, and described in detail in part iii of this work.

With these concepts in mind and tools in hand, we can begin the endeavour of exploring the world of femtochemistry.

1.2 Outline

The thesis will try to present answers to the questions presented in the previous section in the manner described below:

Chapter 2 Describes the molecular systems which have been investigated, all of which contain transition metals.

Chapter 3 Investigates a range of possible methods for approximating relativistic effects which play a role in the structure of molecules containing heavy atoms. In this chapter, a rudimentary DFT introduction is also provided, but can be skipped by readers familiar to the theory.

Chapter 4 Delves into the anchoring of simulation in experiment, by looking at how one calculates the experimental X-ray scattering signal from molecular simulations, and delineates the limits of simulations based on purely classical force-field methods.

Chapter 5 Introduces the QM/MM MD method, the QM code used, and describes how to set up a QM/MM MD simulation in the developed framework.

Chapter 6 Uses QM/MM MD simulations to investigate the structural dynamics of an iridium dimer molecule described in Chap. 2, to uncover new modes, coherence effects and the role of the solvent.

Chapter 7 Presents the current status of QM/MM MD simulations on a bimetallic charge-transfer complex, and compares the resulting solvation effects with the model from Chap. 4.

Chapter 8 Collects the attained results and observations, and presents future goals and ambitions.

The choice of deviating from clumping the theoretical background parts together is deliberate, since reading through it all at once without harvesting a few results in between could be somewhat…dull. This entails that a few results obtained with the electronic structure method described in Chap. 5 are presented in Chap. 3, but the interpretation of those results does not strictly require any of the information given in Chap. 5, so the author believes no real harm is done this way. Conversely, the descriptions of all the molecular systems benefit from being grouped together. This organization could possibly require revisiting Chap. 2 while reading the rest of the chapters, however.

References

1. The Nobel Prize in Chemistry (1999). http://www.nobelprize.org/nobel_prizes/chemistry/. Accessed 1 Oct 2014
2. T.S. Rose, M.J. Rosker, A.H. Zewail, J. Chem. Phys. **88**, 6672 (1988)
3. N.F. Scherer, J.L. Knee, D.D. Smith, A.H. Zewail, J. Phys. Chem. **89**, 5141 (1985)
4. M. Dantus, M.J. Rosker, A.H. Zewail, J. Chem. Phys. **87**, 2395 (1987)
5. D.J. Tannor, *Introduction to Quantum Mechanics: A Time Dependent Perspective* (University Science Books, Sausalito, 2006). ISBN 978-1891389238
6. J.H. Jensen, *Molecular Modeling Basics* (CRC Press, London, 2010)

7. F. Jensen, *Introduction to Computational Chemistry*, 2nd edn. (Wiley, New York, 2007)
8. B. Valeur, *Molecular Fluorescence* (Wiley, New York, 2002). ISBN 3-527-29919-X
9. M. Kasha, Discuss. Faraday Soc. **9**, 14 (1950)

Chapter 2
The Systems of This Project

There are two overall groups of complexes in this study: Members of the d^8-metal complexes, known for the debated nature of the 'd^8–d^8-interactions' [1], that can cause oligomerisation [2, 3]. This entails having two (or more) electron-rich elements close to each other, but weaker bound than in the covalent case, such that large changes in molecular geometry are very likely to occur if the system is perturbed; an opportune model system for dynamics measurable with the XDS-pump-probe method.

The other general category also involves transition metals, but in this case they appear in complexes exhibiting octahedral symmetry, where the metal atom is centered between 6 nitrogen atoms of the ligands. The complexes are of interest both due to an intricate relationship between their spin-configurations and the resulting geometries, and furthermore due to the effect this relationship on their excited-state lifetimes.

2.1 Tetracyanoplatinate, $[Pt(CN)_4]^{2-}$

The origin of the interest in d^8-metal compounds has been reviewed by Harvey [4], focusing on the origin of—and motivation for—the ligands. Gliemann and Yersin [3] provides an in-depth overview of a single member of the square planar complexes: Tetracyanoplatinate (II), $[Pt(CN)_4]^{2-}$ (see Fig. 2.1). According to Harvey, the interest in these compounds originated from the work done by Gray et al. from 1977–1981 [5–8], realising that the $[Rh(CNR)_4]^+$ complex oligomerizes, and that the oligomer structure is concentration dependent. This oligomerization was also found to take place for $[Pt(CN)_4]^{2-}$ ions in solution [2]. Apart from the fact that the metal-metal distance is concentration (-and pressure [9]) dependent, the metal-metal distance can also be varied by changing the counter ion used in the crystallization. Actually, the Pt–Pt distances in the crystals vary with up to 0.7 Å, depending on the counter ion [3].

A.O. Dohn, *Transient Changes in Molecular Geometries and How to Model Them*, Springer Theses, DOI 10.1007/978-3-319-18747-1_2

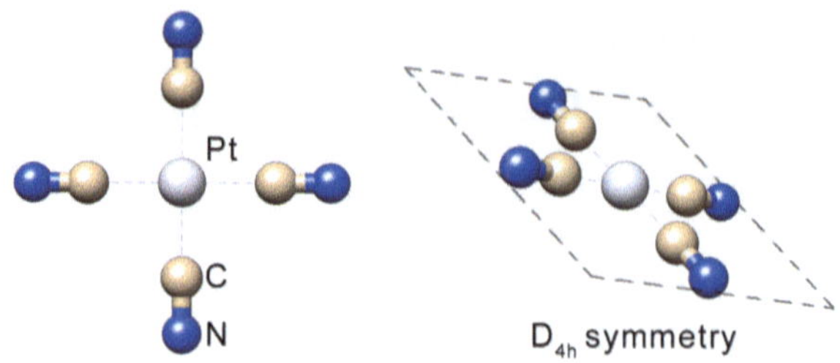

Fig. 2.1 Illustration of the Tetracyanoplatinate complex $[Pt(CN)_4]^{2-}$. The Pt atom is centered between the 4 cyano groups in D_{4h} symmetry, so the complex is *square* planar

The compound can be modeled as a quasi one-dimensional metal-metal bonded chain, exhibiting extremely anisotropic conductivity [10], making it interesting for transport studies and molecular electronics.

This kind of solvated species constitutes a complex system, with oligomers of different numbers of units, and different structures present in the solution at once. Thus, the host of dependencies diminishes the level of attainable control desired for detailed dynamics studies, or later applicability in energy conversion. However, the square planar symmetry of $[Pt(CN)_4]^{2-}$makes it feasible for full, 4-component relativistic QM geometry relaxations, as the symmetry-induced degeneracy can be used to reduce the number of needed computations.

2.2 $M_2(dimen)_4^{2+}$

The need for better control of the binuclear complex structure motivated the search for bridging ligands that could provide this. The most widely employed bridging ligands are shown in Fig. 2.2.

The dimen molecules (see Fig. 2.2, center) have intermediate flexibility when used as ligands in the complex, since they are longer than the 'bridge' (Fig. 2.2, left), but are stabilized by the ring, compared to 'TMB' (Fig. 2.2, right). This makes them the best candidates for obtaining a useful combination of control and flexibility, to mitigate the difficulties in the experimental detection of structural change upon

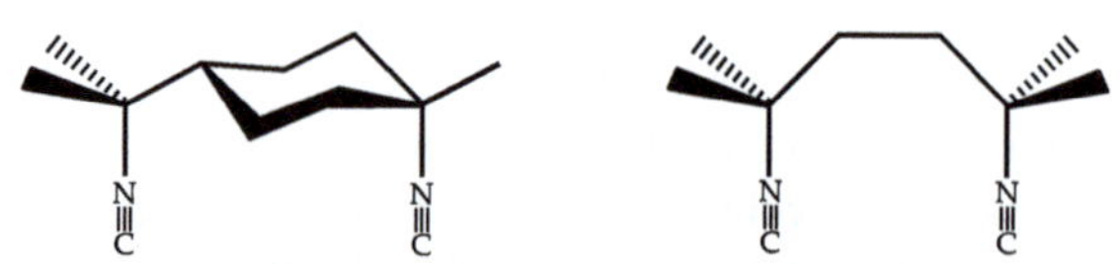

Fig. 2.2 Ligands for the binuclear systems. The complete complexes are made from 2 metal ions, attached at the triple bonded carbon to four opposing ligand units (see Fig. 2.3). Here, the ligands are displayed with increasing flexibility from *left* to *right*

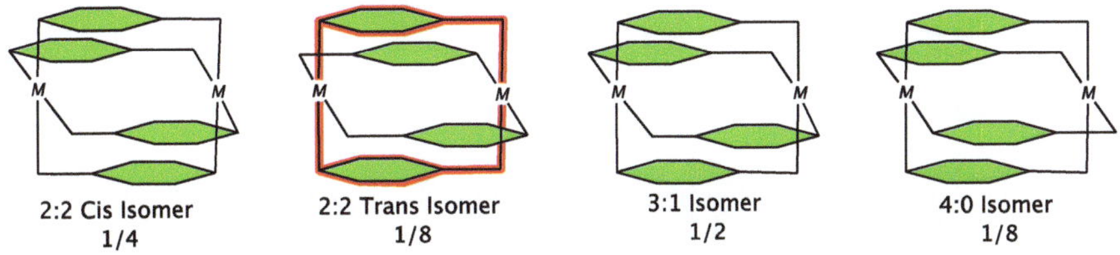

Fig. 2.3 Schematic of the dimen orientation in the four $M_2(dimen)_4^{2+}$ isomers. The *highlighted* part of the 2:2 Trans isomer is represented in the schematic in Fig. 2.4. The statistical distribution in solution [11] is displayed below each isomer. Adapted from [12]

outside action. The cyclohexane ring breaks the ligand symmetry such that the final complex exists in four isomeric structures, as shown in Fig. 2.3.

The Ground State (GS) crystal structure of complexes from this group with the dimen ligand was first determined with M=Rh by Mann [13]. They were dubbed 'windmill'-like in their arrangement (see Figs. 2.3 and 2.4). The suggestion that the structure was similar for all complexes in the group was later confirmed for $Rh_2(dimen)_4^{2+}$ [14] and $[Ir_2(dimen)_4]^{2+}$ as well [11]. Figure 2.4 shows the structure of the $[Ir_2(dimen)_4]^{2+}$ complex, without hydrogens, for clarity. While our group has been experimentally investigating both $[Rh_2(dimen)_4]^{2+}$ [16, 17] and $[Ir_2(dimen)_4]^{2+}$ [18], the Ir variant was chosen for both the further theoretical modelling, and for experimental investigation at an XFEL. This choice is based upon the following reasons:

- The scattering signal from the Ir-containing complex is larger, due to the larger amount of electrons in Ir.
- The excitation fraction of $[Ir_2(dimen)_4]^{2+}$ is higher for the employed experimental setup [17].

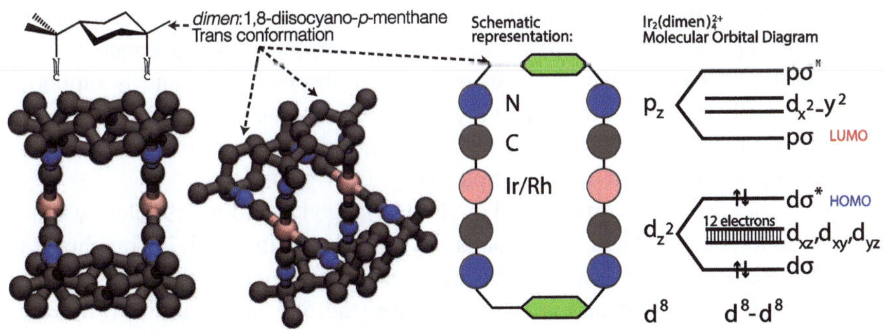

Fig. 2.4 *Left* The 'windmill-like' structure of the $[M_2(dimen)_4]^{2+}$ complex, here shown without hydrogens for clarity. *Center* Schematic representation which only needs to depict half of the complex, due to its symmetry (see Fig. 2.3). The schematic emphasises the atoms of the 'ligand legs' connected to the metals. This representation will be used in the explanation of the simulated dynamics in Chap. 6. *Right* Molecular orbital diagram, reconstructed from [15], featuring the metal-centered, bonding LUMO and the antibonding HOMO

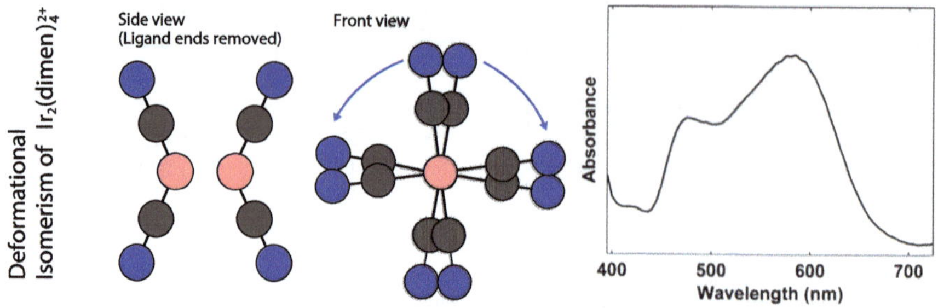

Fig. 2.5 *Left* Deformational isomerism in $[Ir_2(dimen)_4]^{2+}$. The 'Short and Twisted' conformer can only be theoretically described by mapping energy surfaces of the Ir–Ir distance and the twisting angle between the metal ligand planes, while allowing the rest of the structure to relax at each point. *Right* Absorption spectrum of $[Ir_2(dimen)_4]^{2+}$ in acetonitrile: Pumping the molecule at 475 or 585 nm selectively excites the long and eclipsed, or the short and twisted conformer, respectively [19]

- Transient optical spectroscopy has shown coherent oscillations post excitation of this complex [19] (see Fig. 2.7 and accompanying text).

The electronic configuration of this complex features a σ-antibonding[1] Occupied Molecular Orbital (HOMO), while the Molecular Orbital (LUMO) is σ-bonding [15, 20]. Thus, when photoexcited, the complex effectively forms a chemical bond between the two metals. This, in combination with the previously discussed flexibility of the dimen ligand, causes the complex to undergo large structural changes. We have previously reported a 1.3 Å contraction along the Ir–Ir axis for $[Ir_2(dimen)_4]^{2+}$ in solution [18], using pulsed synchrotron radiation. While this contraction is far from unique to the Ir-variant (a solid-phase Rh–Rh contraction of 0.86 Å has also previously been reported [21]), it is, to our knowledge, the largest.

Hartsock et al. [19] have carried out transient spectroscopy measurements on the $[Ir_2(dimen)_4]^{2+}$ complex, and report a deformational isomerism which effectively splits the GS population in two main structures: a 'short and twisted' (Ir–Ir distance of ~3.6 Å, 17° twist, see Fig. 2.5), and a 'long–eclipsed' (Ir–Ir distance of ~4.4 Å, 0° twist). This isomerism has previously been observed [22], and is supported computationally by constrained mappings of the energy landscapes, made in-house by van Driel [17] (see Fig. 2.6) and elsewhere [23], and is reflected in the double-peak in the absorption spectrum of the complex (Fig. 2.5, right).

The transient spectroscopy carried out by Harlang [12] and in [19] clearly shows signs of coherent vibrations. The results are reproduced in Fig. 2.7. The choice of which GS deformational isomer structure is excited influences the ES dynamics: When pumping the short and twisted structure (the 585 nm peak), two dynamic modes are observed, with frequencies of 80 and 119 cm^{-1}, and assigned to a pinch along the Ir–Ir axis and a dihedral twist, respectively. When pumping the long–

[1] A term originating from orbital- and hybridisation theory, where the overlap of atomic s orbitals makes a molecular σ orbital.

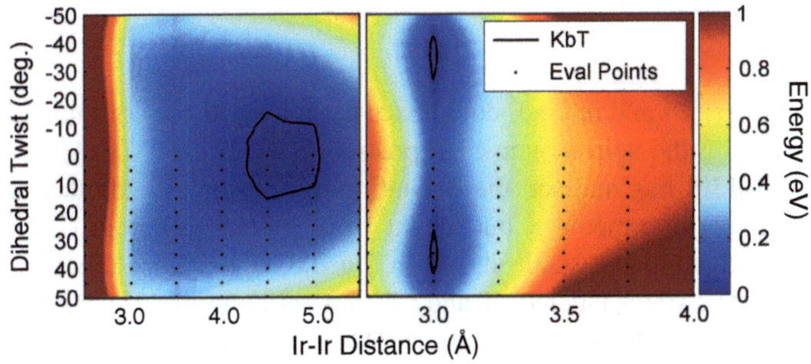

Fig. 2.6 Ground- and excited-state energies (*left* and *right*, respectively) of the 3 : 1 isomer of $[Ir_2(dimen)_4]^{2+}$, evaluated in vacuum as a function of Ir–Ir distance and dihedral angle of the opposing ligands. While the GS potential is very shallow, there is an indication of a minimum around 3 Å and 35°. Reproduced with permission from van Driel and Nielsen [17]

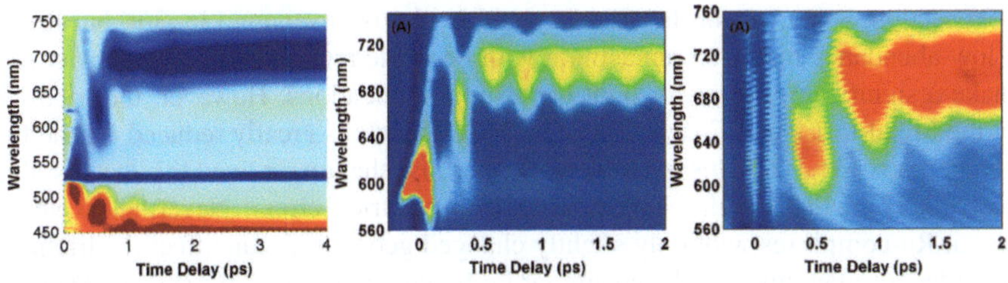

Fig. 2.7 Ultrafast Time-resolved Spectroscopy studies of $[Ir_2(dimen)_4]^{2+}$ in acetonitrile solution. *Left* Transient absorption, pumped at 527 nm [12]. *Middle* Stimulated Emission from pumping at 590 (short and twisted) [19]. *Right* Stimulated Emission pumped at 477 nm (long–eclipsed). Restating the article, the electronic decay of the signal was fitted to a multiexponential decay. The residual between this decay and the signal was Fourier transformed for time delays after $t \geq 0.5$ ps to get a vibrational frequency of 75 cm^{-1}

eclipsed structure (the 475 nm peak), only one mode is observed, with a frequency of 75 cm^{-1}, assigned to the Ir–Ir pinch.

These assignments of the normal mode vibrations can be significantly expanded upon, and more information of the underlying structural dynamics can be revealed through simulations of this system, when anchored in the experimental results. Furthermore, the experiments must of course be carried out in solution, but it is difficult to experimentally distinguish between solvation-induced effects and features purely from internally in the molecule. Simulations can help establish the role of the solvent, and the interplay between the complex and its environment. All this information can then again be used in the interpretation of data gathered at the Linac Coherent Light Source (LCLS) XFEL, where we have attempted to directly probe these coherent motions.

To our knowledge, no attempts (including our own) to freely relax the geometry of the molecule into the short conformer has proved successful, when using DFT methods. This is why we in Chap. 6 focus on simulating the long–eclipsed isomer.

2.3 $[Fe(bpy)_3]^{2+}$

The $[Fe(bpy)_3]^{2+}$- and related- complexes has a long history of interest and controversy regarding the spin dynamics expressed in them [25–32]. As an example, previous work [30] has mapped the excitation-relaxation cascade devoid of transitions to a triplet intermediate, but recent work by collaborators [25] found the relaxation spin dynamics cascade shown in Fig. 2.8, right. Since the relevant occupied orbitals of the states labeled ^3T and $^5T_2$ are antibonding [28], the excitation is accompanied by a Fe–N bond elongation. These dynamics are relevant due to the structure-spin-state interplay, and their effect on the lifetimes of each state. Understanding this interplay, and obtaining control of the lifetimes is important since Fe-complexes like $[Fe(bpy)_3]^{2+}$could hopefully be able to replace their efficient Ru-counterparts in Dye-Sensitized Solar Cell (DSSC)-technologies (see Fig. 2.9 for a schematic of how a DSSC functions). $[Ru(bpy)_3]^{2+}$, described in more detail in the next section, is known for its efficiency within photoconversion [33–36], but Ru is a low-abundance metal, so replacing it with Fe while retaining the efficiency could make a significant impact for light-harvesting applications. However, the lifetime of the Metal-to-Ligand Charge Transfer (MLCT) state is greatly reduced for the Fe counterparts [31, 32]. This is problematic, because the longer lived this state is, the more likely it will be for the charge to enter in an electrical circuit. Nevertheless, studies on Ru-complexes with only slightly changed geometries can affect the lifetime by orders of magnitude [37], and similar Fe-based complexes have shown 100-fold increase in the lifetime, compared to $[Fe(bpy)_3]^{2+}$ [38].

$[Fe(bpy)_3]^{2+}$ has simultaneously obtained the role of a *photophysical factotum* for benchmarking novel tools, techniques, methods, and machinery [25, 31, 32], [VII, I]. This is also the role it will play in this work. The consequences of electronically

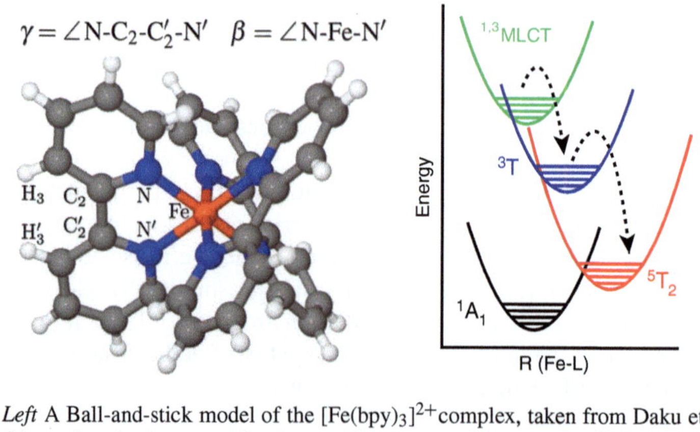

Fig. 2.8 *Left* A Ball-and-stick model of the $[Fe(bpy)_3]^{2+}$complex, taken from Daku et al., using their definitions of angles important for the overall structure of the complex [24]. *Right* Schematic drawing of the states believed to take part in the spin crossover dynamics, as represented in [25], where changes in spin induces changes in Fe-ligand bond length

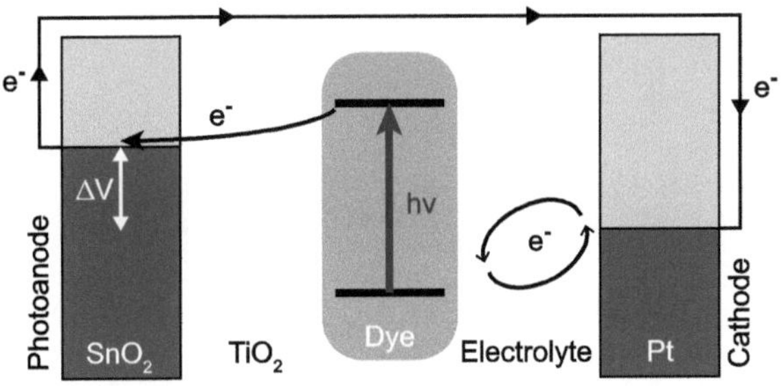

Fig. 2.9 The basic principle of a Dye-Sensitized Solar Cell (DSSC). The semiconductor material is photosensitized by a dye adsorbed to the surface, which, when photoexcited, injects an electron into it, into the anode. This oxidizes the dye, which then in turn oxidizes the electrolyte mediator, which is then reduced by the cathode

exciting $[Fe(bpy)_3]^{2+}$ has already been simulated in the inspiring work by Daku and Hauser [24], where the authors use ab initio MD, an electronic structure-dynamics method akin to Direct Dynamics used in the main part of this project, to show that the Fe–N bond expansion results in further structural deformations. The changes in both electronic and geometric structure causes a change in solvent orientation, and ultimately leads to an expulsion of two water molecules from the first solvation shell. As such, the system of solvated $[Fe(bpy)_3]^{2+}$ exhibits both non-specific, and specific solvation dynamics.

Chapter 4 deals with how solvent-contributions to the scattering signal are simulated, and presents a 'first approximation'-model, based on sampling solvent configurations of fixed-structure solutes using classical MD, and calculating the scattering. This is computationally less demanding than the QM/MM MD, but also has limitations with regards to the level of intricacy of the systems and processes it can describe.

2.4 $[(bpy)_2Ru^{II}(tpphz)Co^{III}(bpy)_2]^{5+}$: Ru=Co

The $[(bpy)_2Ru^{II}(tpphz)Co^{III}(bpy)_2]^{5+}$(Ru=Co, for brevity) system is a member of a larger group of heteronuclear, charge-transfer model complexes. The system can be thought of as a $[Ru(bpy)_3]^{2+}$complex,[2] linked to a $[Co(bpy)_3]^{3+}$complex by an aromatic bridge (Fig. 2.10). The $[Ru(bpy)_3]^{2+}$end is very well-studied [40, 41], in part due to its photo-sensitizing/-conversion abilities [33–36], related to its MLCT-state following electronic excitation [40, 42–48]. The singlet ^{1}MLCT undergoes efficient ISC to a triplet ^{3}MLCT[44, 46], with a lifetime of 600 ns [45].

It has been known for a long time that the metal-ligand bond lengths of Co are changed by 10–45 pm upon ET [49]. Like with $[Fe(bpy)_3]^{2+}$, population of the

[2] Structurally similar to the Fe variant from the last section.

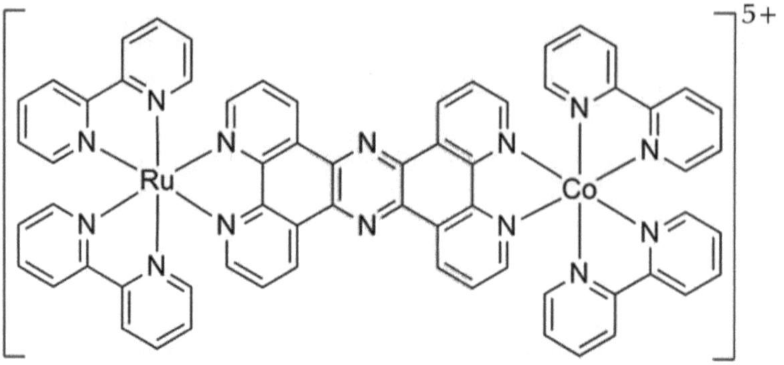

Fig. 2.10 The $[(bpy)_2Ru^{II}(tpphz)Co^{III}(bpy)_2]^{5+}$complex, where (tpphz) = tetrapyrido (3,2-a:2′3′-c:3″,2″-h::2‴,3‴ – j) phenazine [39]. For brevity, the molecule is called Ru=Co throughout this work

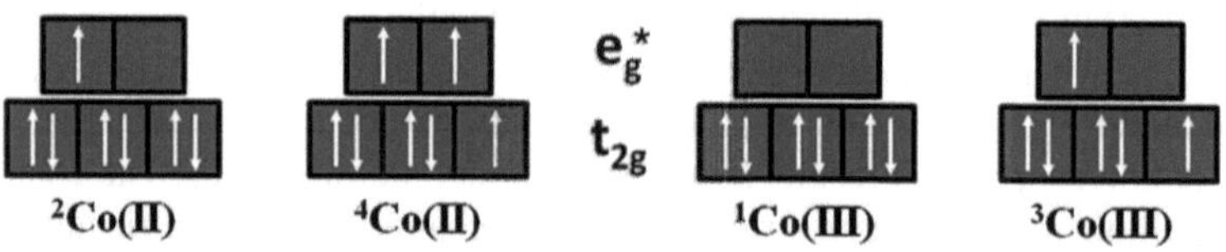

Fig. 2.11 The various electronic valence configurations of Co-systems for their common multiplicities and oxidation states

antibonding e_g^* orbitals (see Fig. 2.11), affects the bond length. We recently experimentally observed that photoexcitation of Ru=Co induced ET from the Ru centre to the Co site, This was accompanied by a spin-flip of ^{2}Co to ^{4}Co, and an on average 0.2 Å elongation of the Co-N bonds in Ru=Co [V]. The spin state of the entire complex when the Co-centre is a doublet, or quartet, is denominated Low Spin (LS), or High Spin (HS), respectively.

The bond-elongation is effectively a relocation of the nitrogen atoms in the Co-centre. Therefore, the reorganization energy—both from solvent- and the intramolecular—will play a large role in the final ET rates of these compounds. These ET rates (both from Ru to Co and the return ET) have been studied for similar complexes [39, 50], where the later article introduces an ISC step to a ^{4}Co(II) state in the ET mechanism, and shows that the return ET in the Ru(III) $\rightarrow$Ru(II) recovery in the complexes is faster for ^{2}Co(II) ions than for ^{4}Co(II), due to a suspected smaller intramolecular reorganization for the doublet pathway [39].

Recently, Co(II)/Co(III) have been proposed as an alternative to the more volatile I_3^-/I^- redox mediators [51, 52] for DSSC. Unfortunately, the same studies have shown that the efficiency drops dramatically. In a combined computational/experimental study, Mosconi et al. [53] argued that the cause of the reduced efficiency was the possible formation of a flexible complex between the cobalt electrolyte and the dye, which brings the cobalt molecule too close to the semiconductor surface.

This allows the cobalt molecule to intercept the semiconductor-injected charges, effectively 'short-circuiting' the DSSC.

In any case, the relevance of these systems are not limited to solar energy conversion; another application perspective is the known (photo-)catalytic activity of cobalt containing complexes in water-splitting processes [54–56].

Apart from the already published results [V] that confirms the Co-N elongation of the final excited state: $[^2Ru^{III}={}^4Co^{II}]$, the first round of results we obtained at the Japanese XFEL SACLA further suggests an intermediate spin-state in the electronic relaxation cascade, before the final state is reached: $[^2Ru^{III}={}^2Co^{II}]$, [III]. The next step in the ongoing investigation, also encompassing experimental data from LCLS, will require simulations of the structural dynamics, and solvent interplay of all these three electronic configurations: the GS, the low spin ES, and the high spin ES. Therefore, Chap. 7 is dedicated to these simulations, where we attempt to achieve a clearer picture of the average solvated structures, their distributions, and how the solvent responds to the charge transfer.

2.5 Summarising the Systems

The systems studied here are interesting in their own right, some as model systems for processes with application-potential within the fields of green technologies and catalysis, others maybe as direct candidates for application in technology. However, it is through the study of the processes themselves—*exemplified* through these systems—where basic knowledge about natural phenomena—and nature itself—can be obtained. Therefore, in this work, the described complexes has been used to concretise the questions posed in the introduction further, in the following ways:

$[Pt(CN)_4]^{2-}$: *What are the options for treating relativistic effects in a satisfactory manner, while ending up with a computationally feasible description for the later studies?*

$[Fe(bpy)_3]^{2+}$: *How far can we get with purely classical methods and steady-state approximations? Furthermore, how does one calculate the experimental signal from simulations?*

$[Ir_2(dimen)_4]^{2+}$: *What dynamic modes channels the dissipation of excess excitation energy? For how long is the process coherent, and what is the cause for decoherence? What is the role of the solvent in this case?*

$[(bpy)_2Ru^{II}(tpphz)Co^{III}(bpy)_2]^{5+}$: *How does the solvent affect the structure and the ISC-induced structural changes? Are the solvation effects purely non-specific, or does this complex exhibit specific solvation dynamics?*

The next part of the thesis will attempt to answer the first two set of questions, to build the foundation for using Direct Dynamics in tackling the final sets of challenges.

References

1. S. Grimme, J. Djukic, Inorg. Chem. **50**, 2619 (2011)
2. J.W. Schindler, R.C. Fukuda, A.W. Adamson, J. Am. Chem. Soc. **104**, 3596 (1982)
3. G. Gliemann, H. Yersin, Struct. Bond. **62**, 87 (1985)
4. P.D. Harvey, Coord. Chem. Rev. **219**, 17 (2001)
5. H.B. Gray, A.W. Maverick, Science **214**, 1201 (1981)
6. I.S. Sigal, K.R. Mann, H.B. Gray, J. Am. Chem. Soc. **102**, 7252 (1980)
7. V.M. Miskowski, I.S. Sigal, K.R. Mann, H.B. Gray, S.J. Milder, G.S. Hammond, P.R. Ryason, J. Am. Chem. Soc. **101**, 4383 (1979)
8. K.R. Mann, N.S. Lewis, V.M. Miskowski, D.K. Erwin, G.S. Hammond, H.B. Gray, J. Am. Chem. Soc. **99**, 5525 (1977)
9. A. Lechner, G. Gliemann, J. Am. Chem. Soc. **111**, 7469 (1989)
10. L. Interrante, *ACS Symposium* (American Chemical Society, Washington DC, 1974)
11. A.G. Sykes, K.R. Mann, J. Am. Chem. Soc. **112**, 7247 (1990)
12. T. Harlang, N. Harrit, *Spectroscopic and X-ray Structural Investigations of* $[Rh_2(\mathrm{dimen})_4]^{2+}$ *and* $[Ir_2(dimen)_4]^{2+}$ *on Pico- and Femtosecond Timescales* (Denmark, Copenhagen, 2010)
13. K.R. Mann, Cryst. Struct. Commun. **10**, 451 (1981)
14. M.R. Rhodes, K.R. Mann, Inorg. Chem. **23**, 2053 (1984)
15. M.D. Roundhill, H.B. Gray, C. Che, Acc. Chem. Res. **22**, 55 (1989)
16. T.B. van Driel, N. Harrit, *Time-Resolved Structural Analysis of* $Rh_2(dimen)_4^{2+}$ *in Solution* (Denmark, Copenhagen, 2010)
17. T.B. van Driel, M.M. Nielsen, *Time-Resolved X-ray Scattering of Molecules in Solution: Approaching the Molecular Movie* (Lyngby, Denmark, 2014)
18. K. Haldrup, T. Harlang, M. Christensen, A. Dohn, T.B. van Driel, K.S. Kjær, N. Harrit, J. Vibenholt, L. Guerin, M. Wulff et al., Inorg. Chem. **50**, 9329 (2011)
19. R.W. Hartsock, W. Zhang, M.G. Hill, B. Sabat, K.J. Gaffney, J. Phys. Chem. A **115**, 2920 (2011)
20. K.R. Mann, J.G.I. Gordon, H.B. Gray, J. Am. Chem. Soc. **97**, 3553 (1975)
21. P. Coppens, O. Gerlits, I.I. Vorontsov, A. Kovalevsky, Y. Chen, T. Graber, M. Gembicky, I. Novozhilova, Chem. Commun. **19**, 2144 (2004)
22. C.L. Exstrom, D. Britton, K.R. Mann, Inorg. Chem. **35**, 549 (1996)
23. B.M. Hunter, R.M. Villahermosa, C.L. Exstrom, M.G. Hill, K.R. Mann, H.B. Gray, Inorg. Chem. **51**, 6898 (2012)
24. L.M.L. Daku, A. Hauser, J. Phys. Chem. Lett. **1**, 1830 (2010)
25. W. Zhang, R. Alonso-Mori, U. Bergmann, C. Bressler, M. Chollet, A. Galler, W. Gawelda, R.G. Hadt, R.W. Hartsock, T. Kroll et al., Nature **509**, 345 (2014)
26. C. Creutz, M. Chou, T.L. Netzel, M. Okomura, N. Sutin, J. Am. Chem. Soc. **102**, 1309 (1980a)
27. J.K. McCusker, K.N. Walda, R.C. Dunn, J.D. Simon, D. Magde, D.N. Hendrickson, J. Am. Chem. Soc. **115**, 298 (1993)
28. P. Gütlich, H.A. Goodwin, Top. Curr. Chem. **233**, 1 (2004)
29. W. Gawelda, A. Cannizzo, V.-T. Pham, F. van Mourik, C. Bressler, M. Chergui, J. Am. Chem. Soc. **129**, 8199 (2007)
30. C. Bressler, C. Milne, V.-T. Pham, A. El-Nahhas, R.M. van der Veen, W. Gawelda, S. Johnson, P. Beaud, D. Grolimund, M. Kaiser et al., Science **323**, 489 (2008)
31. N. Huse, H. Cho, K. Kong, L. Jamula, F.M.F. de Groot, T.K. Kim, J.K. McCusker, R.W. Schoenlein, J. Phys. Chem. Lett. **2**, 880 (2011)
32. H.T. Lemke, C. Bressler, L.X. Chen, D.M. Fritz, K.J. Gaffney, A. Galler, W. Gawelda, K. Haldrup, R.W. Hartsock, H. Ihee et al., J. Phys. Chem. A **117**, 735 (2013)
33. K. Kalyanasundaram, Coord. Chem. Rev. **46**, 159 (1982)
34. K. Kalyanasundaram, M. Grätzel, Coord. Chem. Rev. **77**, 347 (1998)
35. B. O'Reagan, M. Gräzel, Nature **353**, 737 (1991)
36. C. Creutz, M. Chou, T. L. Netzel, M. Okumura, N. Sutin, J. Am. Chem. Soc. **102**, 1309 (1980b)

37. D.G. Brown, N. Sanguatrakun, B. Schulze, U.S. Schubert, C.P. Berlinguette, J. Am. Chem. Soc. **134**, 12354 (2012)
38. L.A. Fredin, M. Pápai, E. Rozsályi, G. Vankó, K. Wärnmark, V. Sundströn, P. Persson, J. Phys. Chem. Lett. **5**, 2066 (2014)
39. H. Torieda, K. Nozaki, A. Yoshimura, T. Ohno, J. Phys. Chem. A **108**, 4819 (2004)
40. A.T. Yeh, C.V. Shank, J.K. McCusker, Science **289**, 935 (2000)
41. A. Juris, V. Balzani, Coord. Chem. Rev. **84**, 85 (1988)
42. N.H. Damrauer, G. Cerullo, A.V. Yeh, T.R. Boussie, C.V. Shank, J.K. McCusker, Science **275**, 54 (1997)
43. K.D. Demadis, C.M. Hartshorn, T.J. Meyer, Chem. Rev. **101**, 2655 (2001)
44. A.C. Basikhuttan, M. Suzuki, S. Nakashima, T. Okada, J. Am. Chem. Soc. **124**, 8398 (2002)
45. A.N. Tarnovsky, W. Gawelda, M. Johnson, C. Bressler, M. Chergui, J. Phys. Chem. B **110**, 26597 (2006)
46. A. Cannizzo, M. Chergui, C. Bressler, W. Gawelda, G. Zgrablic, F. van Mourik, Angewandte Chemie **45**, 3174 (2006)
47. M. Benfatto, S.D. Longa, K. Hatada, K. Hayakawa, W. Gawelda, C. Bressler, M. Chergui, J. Phys. Chem. B **110**, 14035 (2006)
48. D.A. Hoff, R. Silva, L.G.C. Rego, J. Phys. Chem. C **115**, 15617 (2011)
49. K.H. Schmidt, A. Müller, Inorg. Chem. **14**, 2183 (1975)
50. H. Torieda, K. Nozaki, A. Yoshimura, T. Ohno, J. Phys. Chem. A **106**, 11034 (2002)
51. Y. Liu, J.R. Jennings, Y. Huang, Q. Wang, S.M. Zakeeruddin, M. Grätzel, J. Phys. Chem. C **114**, 18847 (2011)
52. H. Nusbaumer, S.M. Zakeeruddin, J.-E. Moser, M. Grätzel, Chem. Eur. J. **9**, 3756–3763 (2003)
53. E. Mosconi, J.-H. Yum, F. Kessler, C.J.G. García, C. Zuccaccia, A. Cinti, M.K. Nazeeruddin, M. Grätzel, F.D. Angelis, J. Am. Chem. Soc. **134**, 19438 (2012)
54. V. Artero, M. Chavarot-Kerlidou, M. Fontecave, Angewandte Chemie **50**, 7238 (2011)
55. D. Shevchenko, M.F. Anderlund, A. Thapper, S. Styring, Energy Environ. Sci. **4**, 1284 (2011)
56. D. Hong, J. Jung, J. Park, Y. Yamada, T. Suenobu, Y. Lee, W. Nam, S. Fukuzumi, Energy Environ. Sci. **5**, 7606 (2012)

Part II
Preliminary Studies

Benchmarking models, testing implementations and limits of approximations. Setting the stage for Direct Dynamics.

Chapter 3
Treating Relativistic Effects in Transition Metal Complexes

Working with transition metal complexes in a computational environment presents itself with a set of added considerations: In force-field methods, the more universal force fields are often not optimized to work with the more exotic elements [1, 2], and (standard) force field methods cannot describe processes such as chemical reactions [3]. In using ab initio methods, problems can arise in how the attractive London dispersion interactions are modelled, as will be touched upon during next part of this work. This chapter, however, is devoted to examining another feature of the electronic structure that becomes relevant when dealing with transition metal complexes: How to describe relativistic changes to the orbital shapes, kinetic repulsion, and other effects that might change the overall geometry of the complex.

The following introduction is based upon [4].

3.1 The Dirac Equation, and the Pauli- and ZORA Hamiltonians

The time-dependent Schrödinger equation, using a non-relativistic, one-electron Hamiltonian[1]:

$$\left(-\frac{1}{2m}\nabla^2 + \mathrm{V}_{ext}\right)|\Psi(\mathbf{r},t)\rangle = i\frac{\partial}{\partial t}|\Psi(\mathbf{r},t)\rangle \tag{3.1}$$

—is not Lorentz-invariant. The Lorentz transformation from one inertial frame to another requires equivalent space and time coordinate, such that the speed of light remains unchanged in all frames. However, the wave function that describes the system is doubly differentiated w.r.t. space and singly differentiated w.r.t. time, and as such, using (methods based on) the Schrödinger equation for solving the electronic structure of a system leaves out relativistic changes to the electronic density.

[1] The electron mass is explicitly included in this section.

A.O. Dohn, *Transient Changes in Molecular Geometries and How to Model Them*, Springer Theses, DOI 10.1007/978-3-319-18747-1_3

The relativistic effects become apparent as the classical velocities of the particles approach the speed of light [5], and as a rule of thumb, the average classical velocity of an electron in the 1s shell is roughly equal to the atomic number[2] [6], so the further we go in the periodic system, the more we will need a relativistic treatment of the electronic density.

A full relativistic description can be obtained by replacing the non-relativistic Hamilton operator, as done by Dirac:

$$\left[c\boldsymbol{\alpha}\cdot\mathbf{p}+\boldsymbol{\beta}mc^2+V_{ext}\right]|\Psi\rangle = i\frac{\partial}{\partial t}|\Psi\rangle \quad \text{where } |\Psi\rangle = \begin{pmatrix}|\Psi_{L\alpha}\rangle\\ |\Psi_{L\beta}\rangle\\ |\Psi_{S\alpha}\rangle\\ |\Psi_{S\beta}\rangle\end{pmatrix} \tag{3.2}$$

where the momentum operator $\mathbf{p} = -i\nabla$ is of the same order derivative as the derivative of time, making the Dirac equation Lorentz-invariant. $\boldsymbol{\alpha}$ and $\boldsymbol{\beta}$ are 4×4 matrices:

$$\alpha = \begin{pmatrix}0 & \sigma_{x,y,z}\\ \sigma_{x,y,z} & 0\end{pmatrix} \quad \beta = \begin{pmatrix}0 & 0\\ 0 & -2\mathbf{I}\end{pmatrix} \tag{3.3}$$

where $\boldsymbol{\sigma}$ are the 2×2 Pauli spin matrices, and $\mathbf{I}$ is a 2×2 unit matrix. The indices on σ is in Einstein notation, so that $\sigma_{x,y,z}$ means a sum of the three Pauli matrices, shown below:

$$\sigma_x = \begin{pmatrix}0 & 1\\ 1 & 0\end{pmatrix} \quad \sigma_y = \begin{pmatrix}0 & -i\\ i & 0\end{pmatrix} \quad \sigma_z = \begin{pmatrix}1 & 0\\ 0 & 1\end{pmatrix} \tag{3.4}$$

In short, the Dirac equation is somewhat more involved than the Schrödinger equation, and the wave functions that solve it now contain four components, and the spin-property is now inherent in the description, since the $\boldsymbol{\sigma}_{x,y,z}$ matrices can be viewed as spin operators, and two main components $|\Psi_L\rangle$ and $|\Psi_S\rangle$ of the relativistic wave function (termed the 'Large' and the 'Small' component) each has two spin components, α and β. Solving the Dirac equation produces a spectrum described in the left side of Fig. 3.1, which compares it to the non-relativistic result.

We can recover the time-independent Schrödinger equation by eliminating the small component and setting the speed of light $c \to \infty$. First, we write out the time-independent Dirac equation, the Hamiltonian of which is often called the Dirac-Coloumb Hamiltonian, $\boldsymbol{H_D}$:

$$\boldsymbol{H_D}|\Psi\rangle = E|\Psi\rangle \Rightarrow$$

$$\left[c\begin{pmatrix}0 & \boldsymbol{\sigma}_{x,y,z}\\ \boldsymbol{\sigma}_{x,y,z} & 0\end{pmatrix}\cdot\boldsymbol{p}+\begin{pmatrix}0 & 0\\ 0 & -2\boldsymbol{I}\end{pmatrix}mc^2+V_{ext}\right]\begin{pmatrix}|\Psi_L\rangle\\ |\Psi_S\rangle\end{pmatrix} = E\begin{pmatrix}|\Psi_L\rangle\\ |\Psi_S\rangle\end{pmatrix} \tag{3.5}$$

[2]This is a rough estimation, based on the energy of a 1s electron in hydrogen in atomic units is $V = -\frac{-Z^2}{2}$, and that the classical virial theorem: $T = \frac{v^2}{2} = -\frac{Z^2}{2}$.

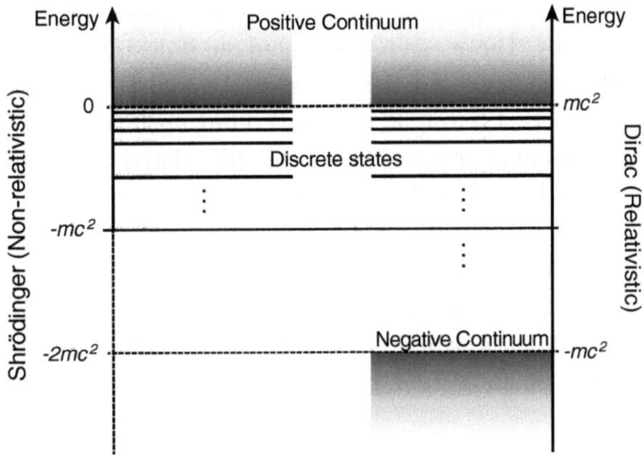

Fig. 3.1 Scheme illustrating the differences of the spectra of the non-relativistic and relativistic Hamiltonian. By solving the Dirac equation, apart from a shift in zero-point for the energy, one also obtains a negative continuum of unbound states

Then, we factor out the Dirac equation into two equations from the block matrices (and omitting the indices on $\boldsymbol{\sigma}$ for brevity):

$$c(\boldsymbol{\sigma} \cdot \boldsymbol{p})|\Psi_S\rangle + V_{ext}|\Psi_L\rangle = E|\Psi_L\rangle \tag{3.6}$$

$$c(\boldsymbol{\sigma} \cdot \boldsymbol{p})|\Psi_L\rangle + (-2mc^2 + V_{ext})|\Psi_S\rangle = E|\Psi_S\rangle \tag{3.7}$$

The small component can be isolated in the second equation:

$$\begin{aligned}(2mc^2 + E - V_{ext})|\Psi_S\rangle &= c(\boldsymbol{\sigma} \cdot \boldsymbol{p})|\Psi_L\rangle \\ |\Psi_S\rangle &= \frac{c(\boldsymbol{\sigma} \cdot \boldsymbol{p})}{(2mc^2 + E - V_{ext})}|\Psi_L\rangle \\ &= K\frac{(\boldsymbol{\sigma} \cdot \boldsymbol{p})}{2mc}|\Psi_L\rangle \end{aligned} \tag{3.8}$$

where $K = \left(1 + \frac{E-V_{ext}}{2mc^2}\right)^{-1}$. If this expression for $|\Psi_S\rangle$ is inserted in to Eq. 3.6, we get

$$\left[\frac{1}{2m}(\boldsymbol{\sigma} \cdot \boldsymbol{p})K(\boldsymbol{\sigma} \cdot \boldsymbol{p}) + (V_{ext} - E)\right]|\Psi_L\rangle = 0 \tag{3.9}$$

and as $c \to \infty$, $K \to 1$, due to the way the Pauli matrices are constructed, the $\boldsymbol{\sigma}$'s cancel out [4], we will end up with:

$$\left[\frac{\boldsymbol{p}^2}{2m} + V_{ext}\right]|\Psi_L\rangle = E|\Psi_L\rangle \tag{3.10}$$

or, the time-independent Schrödinger equation, but for the two-component spin wave functions. These can be separated out if desired, since the momentum operator does not contain any spin dependence.

Instead of setting $K = 1$ and ending up almost where we started, we can make a 2-component approximation to the full relativistic description by the series expansion for small x values $1/(1 + x) = 1 - x + x^2 - x^3 \ldots$, and only keeping the first two terms, such that $K \approx 1 - \frac{E-V_{ext}}{2mc^2}$, which holds in most cases (but not all, as we shall see), since the denominator contains c^2. After a considerable amount of algebra (which is skipped here, but more info can be found in [4, 5, 7]) obtain the Pauli equation, where the often-mentioned 'relativistic effects' are neatly defined as relativistic correction terms to the non-relativistic Hamiltonian:

$$\left[\underbrace{\frac{\boldsymbol{p}^2}{2m} + V_{ext}}_{\substack{\text{Non-Relativistic}\\ \text{Terms}}} + \underbrace{-\frac{\boldsymbol{p}^4}{8m^3c^2}}_{\substack{\text{Mass-Velocity}\\ \text{Correction}}} + \underbrace{\frac{Z\boldsymbol{s}\cdot\boldsymbol{L}}{2m^2c^2r^3}}_{\substack{\text{Spin-Orbit}\\ \text{Term}}} + \underbrace{\frac{Z\pi\delta(\boldsymbol{r})}{2m^2c^2}}_{\substack{\text{Darwin}\\ \text{Correction}}} \right] |\Psi_L\rangle = E|\Psi_L\rangle \tag{3.11}$$

The Mass-Velocity term arises from the dependence of the electron mass on its velocity. The spin-orbit term couples the orbital angular momentum of the electron with its spin, and the Darwin correction can be interpreted as the electron making a high-frequency oscillation around its mean position. The mass-velocity- and Darwin-correction terms are often named 'scalar relativistic' effects. We mention here in passing, that in principle, the ISC processes described in the introduction would not be allowed if the spin-orbit coupling shown in the above did not exist.

The expansion $K \approx 1 - \frac{E-V_{ext}}{2mc^2}$ is only valid for $E - V \ll 2mc^2$, but close to the nuclei, the external potential will go towards $-\infty$ such that the expansion becomes invalid. Instead, one can define

$$K' = \left(1 + \frac{E}{2mc^2 - V_{ext}}\right)^{-1} \tag{3.12}$$

which will have the first term of the expansion $K' \approx 1 - \frac{E}{2mc^2 - V_{ext}}$ and following the same strategy as before, one will (eventually) arrive at the Zeroth Order Regular Approximation (ZORA) Hamiltonian [8–11]:

$$\left[\frac{c^2\boldsymbol{p}^2}{2mc^2 - V_{ext}} + \frac{2c^2}{(2mc^2 - V_{ext})^2} - \frac{Z\boldsymbol{s}\cdot\boldsymbol{L}}{r^3} + V_{ext}\right] |\Psi_L\rangle = E|\Psi_L\rangle \tag{3.13}$$

which, contrary to the Pauli equation, is variationally stable [5], which means that it is useful in the traditional energy minimization framework, even though the non-relativistic, scalar relativistic and spin-orbit terms are not as neatly distinguished in this approximation. The ZORA Hamiltonian has been implemented in the ORCA program, and presents itself as a way of approximating the full, 4 component relativistic description while saving a considerable amount of computational cost.

3.2 Effective Core Potentials/Pseudopotentials and PAW

In order to save computational effort, systems containing many electrons are often approximated by replacing the core electrons of the atoms in the system with Effective Core Potentials (ECPs). Coincidentally, this allows for an implicit incorporation of the most important relativistic effects simply by parametrising the ECP with respect to corresponding relativistic, all-electron data [12, 13].

So when generating an ECP for an element, one must first produce or obtain a high-quality, (relativistic) all-electron wave function for an isolated atom. Then, the valence orbitals are replaced by pseudo-orbitals, that are smoother, thus requiring expansions in smaller basis sets (described in the next section), when used. The core electrons must then be replaced by analytically solvable functions of the nuclear-electron distance. The resulting potential should then of course match, or be fitted to match the relativistic behaviour of the core electrons, as well as the original all-electron valence orbitals.

The Projector Augmented Wave method (PAW) method can [14, 15] be considered an advanced pseudopotential technique that retains the core electrons (more info in Sect. 5.1, Chap. 5). The Grid based Projector Augmented Wave method (GPAW) technique [16, 17] is somewhat unique compared to the more standard computational chemistry methods, and will be introduced when necessary, but for the time being it is sufficient to think of GPAW as an advanced ECP-method, where the PAW functions are not expanded in either plane waves or a Gaussian basis set, but evaluated on a real space grid, using finite difference methods.

All the different relativistic methods are applicable within DFT, the outlines of which will be briefly introduced in the following section.

3.3 Density Functional Theory

In this section, we briefly open the lid on the black box of DFT codes, to have a brief look of the basic principles behind the theory. However, much more comprehensive descriptions are available [3, 4, 18].

The main idea behind DFT is that the electron density $\rho(\mathbf{r})$ has all the information on the variables of which the Hamiltonian is a function of: The number of electrons, and the potential created by the positions and charges of the nuclei in the system. Thus, the energy of the system can be described as a functional of the electron density, i.e. $E[\rho(\mathbf{r})]$, which has the general form:

$$E[\rho(\mathbf{r})] = \int \rho(\mathbf{r}) \frac{\partial E[\rho(\mathbf{r})]}{\partial \rho(\mathbf{r})} d\mathbf{r} \tag{3.14}$$

This would reduce the complexity of the problem of electronic structure a great deal, seeing as how a wave function for an N electron system contains $4N$ variables (three spatial and one spin coordinate for each electron), whereas each electronic density only depends on **r**, independent of **N**.

In 1964, Hohenberg and Kohn helped along this idea by proving that (**1**) no two electronic densities will give the same energy (i.e. the total energy is a unique functional of the electron density), and (**2**) that the exact GS density will give the exact GS energy, provided that the exact density functional is used.

To prove (**1**), assume that two different external potentials (which is often the potential from the nuclei, according to the Born-Oppenheimer approximation, but could also contain terms of other external fields), V_{ext} and V'_{ext} give the same $\rho(\mathbf{r})$—i.e. that (**1**) is not true. This means that we also have two different Hamiltonians **H** and **H**$'$ with corresponding wave functions $|\Psi\rangle$ and $|\Psi\rangle'$. The variational principle says that the energy of an approximate wave function must be higher than the exact one, so if we switch up the Hamiltonians and the wave functions and apply the principle to each one, we will arrive at an contradictory result. Here it is shown for the combination of **H** and $|\Psi'\rangle$:

$$\begin{aligned}
\langle\Psi'|\mathbf{H}|\Psi'\rangle &> E_0 \\
\langle\Psi'|\mathbf{H}'|\Psi'\rangle + \langle\Psi'|\mathbf{H} - \mathbf{H}'|\Psi'\rangle &> E_0 \\
E'_0 + \langle\Psi'|\mathbf{V_{ext}} - \mathbf{V'_{ext}}|\Psi'\rangle &> E_0 \\
E'_0 + \int \rho(\mathbf{r})(V_{ext}(\mathbf{r}) - V'_{ext}(\mathbf{r}))d\mathbf{r} &> E_0
\end{aligned} \tag{3.15}$$

and the other combination of **H**$'$ and Ψ will follow the same way:

$$\begin{aligned}
E_0 + \langle\Psi|\mathbf{V'_{ext}} - \mathbf{V_{ext}}|\Psi\rangle &> E'_0 \\
E_0 + \int \rho(\mathbf{r})(V'_{ext}(\mathbf{r}) - V_{ext}(\mathbf{r}))d\mathbf{r} &> E'_0 \\
E_0 - \int \rho(\mathbf{r})(V_{ext}(\mathbf{r}) - V'_{ext}(\mathbf{r}))d\mathbf{r} &> E'_0
\end{aligned} \tag{3.16}$$

which added up with Eq. 3.15 becomes:

$$E_0 + E'_0 > E_0 + E'_0$$

—an impossible inequality. Thus, the assumption that two different external potentials $V_{ext}(\mathbf{r})$ and $V'_{ext}(\mathbf{r})$ can give the same electronic density $\rho(\mathbf{r})$ is wrong. So there is a one-to-one correspondence between the electron density and the potential.

The proof of the second theorem uses the first: Since two different wave functions $|\Psi'\rangle$ and $|\Psi\rangle$ cannot yield the same electronic density $\rho(\mathbf{r})$, the density $\rho(\mathbf{r})'$[3] would be the result of $|\Psi'\rangle$ (and *not* $|\Psi\rangle$). Thus,

$$E[\rho(\mathbf{r})'] = \langle\Psi'|\mathbf{H}|\Psi'\rangle > \langle\Psi|\mathbf{H}|\Psi\rangle = E[\rho(\mathbf{r})] \tag{3.17}$$

[3] Which still integrates up over all space to the number of electrons in the system.

which is recognized as the variational principle, but for electronic densities instead of wave functions.

All this means that, if the exact functional $E[\rho(\mathbf{r})]$ connecting the energy and the density was known, the complexity of obtaining the energy of a quantum mechanical system would be greatly reduced in comparison to wave function approaches.

However, this functional is *not* known.

The reason why DFT was not simply put back into the lowest drawer was given by Kohn and Sham [19], who reintroduced wave functions as single-particle orbitals $|\phi_i\rangle$, describing the kinetic energy of a *non-interacting* system:

$$\mathbf{T}_{\text{nonint}} = -\frac{1}{2m}\sum_i f_i\langle\phi_i|\nabla^2|\phi_i\rangle \tag{3.18}$$

where f_i are the occupation numbers of the orbitals. Therefore, the DFT method uses basis sets like Hartree-Fock, or SCF-theory [4], so the Kohn-Sham orbitals are represented as linear combinations of atomic orbitals:

$$|\phi_i\rangle = \sum_n c_{n,i}|\chi_n\rangle \tag{3.19}$$

where $|\chi_n\rangle$ is often expanded in a(nother) linear combination of Gaussian functions. If only enough functions are used to contain the electrons of the system, the basis set is *minimal*, if twice as many functions are used, it is called *double zeta*, three times as many: *triple zeta*, etc…The 'zeta' term arising from the most often used symbol for the exponent. The larger the amount of functions used in the expansion, the more accurate (and computationally expensive) the basis set description will be.

The exact kinetic energy levels of the non-interacting system are obtained by solving the one-electron Schrödinger (i.e. the KS) equation:

$$\mathbf{H}_{DFT}|\phi_i\rangle = E_i|\phi_i\rangle \tag{3.20}$$

where the electronic density are given through the single particle states:

$$\rho(\mathbf{r}) = \sum_i^{N_e} f_i\langle\phi_i|\phi_i\rangle = \sum_i^{N_e} f_i\,|\phi_i(\mathbf{r})|^2 \tag{3.21}$$

and the general expression for the DFT energy becomes:

$$E_{DFT}[\rho(\mathbf{r})] = T_{\text{nonint}}[\rho(\mathbf{r})] + \int \rho(\mathbf{r})V_{ext}(\mathbf{r})d\mathbf{r} \tag{3.22}$$

However, the electrons *are* of course interacting, and thus the total energy expression for the DFT energy defines the energy originating from electronic interaction in terms of what is missing from the exact expressions (here we leave out the r-dependency of the density, for clarity):

$$E_{\text{DFT}}[\rho] = T[\rho] + E_{ne}[\rho] + J[\rho] + E_{XC}[\rho] \quad (3.23)$$

with

$$E_{XC}[\rho] = (T[\rho] - T_{\text{nonint}}[\rho]) + (E_{ee}[\rho] - J[\rho]) \quad (3.24)$$

$J[\rho]$ is the Coloumb repulsion between electron pairs, $E_{ne}[\rho]$, is the nuclei-electron potential, and E_{XC} is the exchange-correlation term: Basically 'what is left', when reintroducing interactions between electrons. The first parenthesis is often termed the kinetic correlation energy, while the second contains both potential correlation and exchange energy, such that $E_{ee}[\rho]$ is the total, exact potential energy of the many-particle system.

Finally, the same DFT approximations can also be made to the fully relativistic Hamiltonian presented in the previous section, the further details of which are beyond the scope of this work.

There is a host of functionals approximating the unknown, exact functional for the exchange-correlation term. They vary in complexity and suitability for different systems, but they are generally computationally much less expensive than traditional, high-level computational chemistry methods , while still often achieving considerably accurate results. The exchange-correlation functionals employed later in this work, for the Direct Dynamics simulations, are all within the Generalized Gradient Approximation (GGA) group. GGA functionals make exchange and correlation energies dependent on derivatives of the electronic density, as well as the density itself. For these preliminary studies of relativistic effects, hybrid GGA will also be used. These include a fraction of the exchange energy from Hartree-Fock theory, at an added computational cost.

The exchange-correlation functionals can be totally ab initio in nature, i.e. they are fitted to data from higher order methods, and should recover analytic/known results, or they can be fitted to experimental data, which makes them semiempirical. Examples of the first category are P86 [20], PBE [21], VWN [22] and PW91 [23], whereas the semiempirical category includes functionals such as B(88) [24], LYP [25], and more. Thus, labeling e.g. computational work done with the BLYP (B(88) for exchange, LYP for correlation) as ab initio would not be entirely correct.

3.4 Benchmarking the Methods on Tetracyanoplatinate

As mentioned in Chap. 2, $[Pt(CN)_4]^{2-}$ is one of the 'most basic' members of the $d^8 - d^8$ group, and while the simple cyano-ligands do not allow for much structural control, compared to the other studied systems, the smaller amount of atoms in the complex, and its planar (D_{4h}) symmetry makes it feasible to structurally relax the system using the Dirac-Coloumb Hamiltonian within the DFT framework. Thus, this system is optimal for benchmarking the various relativistic approximations described in the previous sections against the fully relativistic result. Pt is the transition metal with the highest atomic number of the metals in the systems of this work, so it is also assumed to be the system with the most pronounced relativistic behaviour.

Earlier work [26] has shown that if solvation is neglected completely, a structural optimization of the $[Pt(CN)_4]^{2-}$ dimer will not converge. This, together with the 'silicon-limits' imposed by the computational effort needed for the fully relativistic geometry relaxations, means that the benchmarking has been carried out on the $[Pt(CN)_4]^{2-}$ monomer, where we compare the optimized Pt–C and C–N bond lengths obtained with the different descriptions.

3.5 Initial Calculations

The initial mapping of the dependence of the structure of the molecule on the employed basis set for a set of commonly used exchange-correlation functionals were carried out before the start of this Ph.D. project, so they are not included here. The interested reader can seek more information in [VI], which was written during this project, and combines these studies with the final calculations, also carried out as a part of this Ph.D. project. Here, we simply recount the conclusion of the basis-set study: With the largest basis sets employed in the different approaches, sufficient basis set convergence has been reached—in particular in the light of the much larger variations with the exchange-correlation functionals, as discussed in the following section.

3.6 Comparing the Methods

Different studies on similar transition metal compounds reported satisfactory results using various exchange-correlation functionals, such as B3LYP [36, 37] and PBE0 [36, 38]. Both are hybrid functionals, but the PBE0 exchange-correlation functional includes 5 % more exact Hartree-Fock exchange than B3LYP, in the standard setup. Here, we compare their performance with their two corresponding non-hybrid functionals BLYP and PBE. Figure 3.2 compares the results for the four functionals and the largest basis sets with all relativistic methods.

With regards to the choice of functional, a significant difference between the change in platinum-carbon and carbon-nitrogen bond length is observed. For the C–N bond length, the crucial factor is whether the functional includes exact Hartree-Fock exchange or not. The Pt–C bond, on the other hand, is in general more sensitive and in particular the choice of correlation functional has a larger impact. Employing the hybrid functionals shortens thus the C–N bond length by 1.2–1.4 pm or approx-

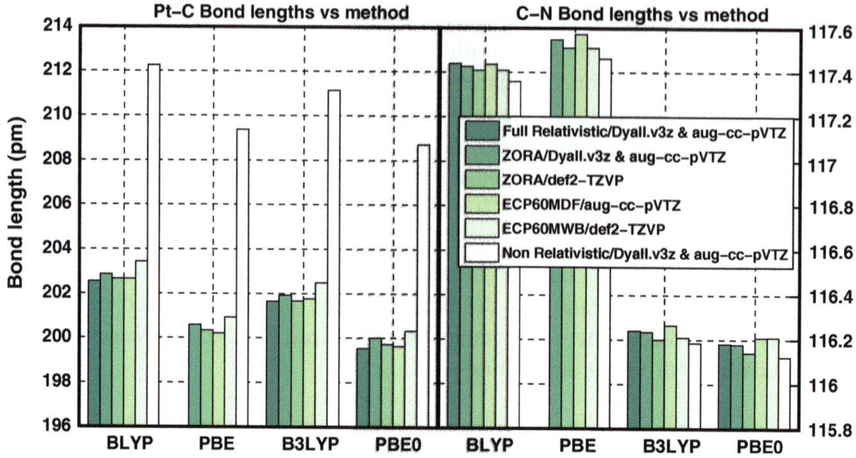

Fig. 3.2 Benchmark of relativistic approximations on the Tetracyanoplatinate geometry. The scale of the Pt–C plot is ten times larger than the C–N plot. Neglecting any type of relativistic treatment has a large impact on the Pt–C bond. The labels 'Dyall.v3z', 'aug-cc-pVTZ', and 'def2-TZVP' do all correspond to various triple-zeta basis sets [27–33], and their corresponding ECP [34, 35]

imately 1% for all methods, while the differences between the exchange-correlation functionals BLYP and PBE or B3LYP and PBE0 amounts to not more than 0.1 pm. For the Pt–C carbon bond employing a hybrid functional shortens the bond length by almost 1 pm or 0.5 %, whereas using PBE or PBE0 instead of BLYP or B3LYP shortens the bond length by 2.0–2.5 pm or 1%.

3.7 The Origin of the Relativistic Pt–C Contraction

Figure 3.2 also compares the bond lengths obtained using various approximations to the fully relativistic 4-component treatment using the Dirac-Coloumb Hamiltonian. The Pt–C bond contracts when increasing the precision of the relativistic description, and becomes shortest when using the Dirac-Coloumb Hamiltonian, for all the functionals in agreement with the previous study on the cationic platinum carbene complex $PtCH2^{+}$ [37], and the general rule-of-thumb that the effect of relativity on chemical bonds from a heavy atom is in most cases a contraction. This might seem counter-intuitive due to the fact that it is the 5d atomic orbitals from Pt that contributes to the Pt–CN bond, orbitals that are known to expand when treated relativistically [6]. However, it has been known for a long time that bond-length contractions are *not* caused by orbital contractions, but rather due to the relaxation of the kinetic repulsion [39, 40]. If a bond is contracted, the electrons will have to occupy a smaller volume, which will result in a rise of kinetic energy, due to the uncertainty principle. However, the relativistic mass-velocity correction becomes more negative with

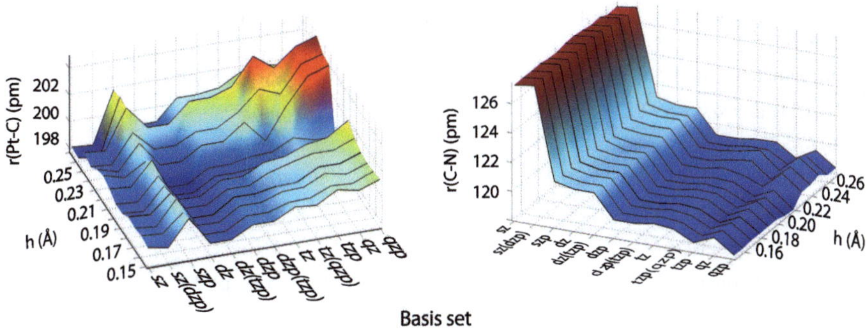

Fig. 3.3 $[Pt(CN)_4]^{2-}$ Bond lengths as a function of real-space grid spacing h, and basis set size. *Left* Pt–C, *Right* C–N. The bold line on both plots represent the value $h = 0.18$ Å, which is often employed in GPAW

increasing non-relativistic kinetic energy, thus diminishing the relativistic kinetic repulsion term, effectively shifting the minimum of the total binding potential. As such, our results are in accordance with the expected effects of a relativistic treatment.

3.8 GPAW Results

While GPAW supports a long list of functionals, we restrict this study to only encompass the PBE functional, which is used in the later QM/MM MD studies. A test of basis set-dependence akin to what was done in [VI] is carried out, but with GPAW, the electronic density is solved on a real-space grid via finite-difference methods,[4] which means the structural convergence now also depends on the gridspacing, h.

The calculations were carried out in a cell of dimensions 16 x 16 x 10 Å, to avoid any wave function cutoff, and the QuasiNewton optimizer from the ASE-package [41] was used to minimize the forces in the structure to a convergence limit of 0.05 eV/Å.

Since there is now two structural convergence parameters instead of only the size of the basis set, we present the data in the surface plots of Fig. 3.3. The figure shows that, for the Pt–C bond, grid-spacings larger than 0.22 Å produce inconsistent results, with distances fluctuating with up to 6 pm when varying the basis set. When using grid spacings below 0.18 Å, basis sets from dzp and up increases the bond length with a maximum of 2 pm, not far from the largest change observed for the initial ECP studies of [VI].

The C–N bond length is much more affected by basis set size than grid spacing, but the description seems to converge around dzp-size. Going from tzp to qzp only changes the bond length by 0.2 pm, for $h = 0.18$ Å.

[4]Details in Sect. 5.1, Chap. 5.

Table 3.1 Deviations from the 4-component results for the optimized Pt–C and C–N bond lengths of $[Pt(CN)_4]^{2-}$ in pm

Approximate method	Δr(Pt–C)/pm	Δr(C–N)/pm
ZORA/Dyall.v3z/cc-pVTZ	0.45	0.05
ZORA/def2-TZVP	0.21	0.01
ECP60MDF/cc-pVTZ	0.21	0.05
ECP60MWB/def2-TZVP	0.79	0.01
GPAW/$h = 0.15$Å-tzp	0.98	1.10

The four-component calculation compared to here is PBE/Dyall.v3z/cc-pVTZ

We can now compare all the relativistic approximations using the PBE functional, to the 4-component calculations of same basis set size, as is done in Table 3.1.

Where all the other approximations do better for the C–N bond, the GPAW method deviates from the 4-component result by an equal amount for both bonds. This indicates that the deviation is not due to the relativistic approximations. All methods come within 1 pm of the fully relativistic bond-length, an error which is easily overshadowed by other changes, such as the ones experienced from switching exchange-correlation functional. All in all, this means that it is very unlikely that later results obtained with GPAW will be contaminated with errors from the description of relativistic effects.

3.9 Comparison to Experiment

Finally, we briefly compare our theoretical predictions of bond lengths with one experimental X-ray and neutron scattering study [42] on -$[K_2Pt(CN)_4Br_{0.30}{\cdot}3.2H_2O]$ crystals. The X-ray and neutron scattering C–N bond lengths are 117.0 and 116.4 pm, respectively, which are in excellent agreement with our calculated results, even though the bromine and crystal water is likely to perturb the bond lengths. The same study also reports Pt–C bond lengths of 200.7 and 200.1 pm, measured using X-rays and neutrons, respectively. These differ by at most only 2 pm from our 4-component results.

3.10 Conclusions

We find, as expected, that relativistic effects contract the Pt–C bond, making it approx 4.6–4.8% shorter (depending on the functional used), when comparing the full relativistic calculation to the non relativistic, while the C–N bond is not affected (as) much. This relativistic bond contraction is almost perfectly reproduced in scalar ZORA calculations and also in calculations with relativistic ECP and in GPAW. The predicted bond lengths are in good agreement with experimental X-ray and neutron scattering values for $[K_2Pt(CN)_4Br_{0.30} \cdot 3.2H_2O]$ crystals.

We observe that addition of exact Hartree-Fock (HF) exchange in the hybrid functionals plays a large role for the C–N bond lengths, but only a small role for the Pt–C bond lengths, whereas it is the choice of correlation functional, which has a large effect on the Pt–C bond lengths with PBE and its hybrid version generally producing shorter Pt–C bond lengths than B(3)LYP. Furthermore, we note that the choice of DFT functional has a larger effect on the bond lengths than the level of relativistic treatment. Finally, we observe that an adequate description of the distance from the heavy Pt atom to the C atom, as expected, requires *some* method of relativistic treatment, but the different levels of approximations used, all show good performance. Thus, the choice of relativistic method to be employed in the main studies of this work should reflect other considerations, such as computational effort, applicability to the Direct-Dynamics framework, and so on. Considering all these factors, we have chosen GPAW, since, as we shall see, it possesses the computational efficiency imperative to the costly direct-dynamics simulations, and the real space method allows for a simple interfacing to regions with molecules described by classical point charges.

References

1. H. Heinz, R.A. Vaia, B.L. Farmer, R.R. Naik, J. Phy. Chem. C **112**, 17281 (2008)
2. Z. Shu, G.J. Davies, Phys. Status Solidi A **78**, 595 (1983)
3. J.H. Jensen, *Molecular Modeling Basics* (CRC Press, 2010)
4. F. Jensen, *Introduction to Computational Chemistry*, 2nd edn. (Wiley, New York, 2007)
5. J. Autschbach, J Chem. Phys. **136**, 150902 (2012)
6. P. Pyykkö, J. Desclaux, Acc. Chem. Res. **12**, 276 (1979)
7. M. Barysz, Y. Ishikawa, *Relativistic Methods for Chemists. Challenges and Advances in Computational Chemistry and Physics* (Springer, 2010)
8. C. Chang, M. Pelissier, P. Durand, Phys. Scr. **34**, 394 (1986)
9. E. van Lenthe, E.J. Baerends, J.G. Snijders, J. Chem. Phys. **99**, 4597 (1993)
10. E. van Lenthe, E.J. Baerends, J.G. Snijders, J. Chem. Phys. **101**, 9783 (1994)
11. E. van Lenthe, E.J. Baerends, J.G. Snijders, J. Chem. Phys. **105**, 6505 (1996)
12. M. Dolg, X. Cao, Chem. Rev. **112**, 403 (2011)
13. X. Cao, M. Dolg, WIREs Comput. Mo. Sci. **1**, 200 (2011)
14. P.E. Blöchl, Phys. Rev. B **50**, 17953 (1994)
15. P.E. Blöchl, C.J. Först, J. Schimpl, Bull. Mater. Sci. **26**, 33 (2003)
16. J. Mortensen, L. Hansen, K.W. Jacobsen, Phys. Rev. B **71**, 035109 (2005)
17. J. Enkovaara, C. Rostgaard, J.J. Mortensen, J. Chen, M. Dulak, L. Ferrighi, J. Gavnholt, C. Glinsvad, V. Haikola, H.A. Hansen et al., J. Phys. **22**, 253202 (2010)
18. J. Kohanoff, *Electronic Structure Calculations for Solids and Molecules* (Cambridge University Press, Cambridge, 2006)
19. W. Kohn, L.J. Sham, Phys. Rev. **140**, 1133 (1965)
20. J.P. Perdew, Phys. Rev. B **33**, 8822 (1986)
21. J.P. Perdew, K. Burke, M. Ernzerhof, Phys. Rev. Lett. **77**, 3865 (1996)
22. S.H. Vosko, L. Wilk, M. Nusair, Can. J. Phys. **58**, 1200 (1980)
23. J.P. Perdew, J.A. Chevary, S.H. Vosko, K.A. Jackson, M.R. Pederson, D.J. Singh, C. Fiolhais, Phys. Rev. B **46**, 6671 (1992)
24. A.D. Becke, Phys. Rev. A **38**, 3098 (1988)
25. C. Lee, W. Yang, R.G. Parr, Phys. Rev. B **37**, 785 (1988)

26. A. Dohn, N. Harrit, *Structural Investigations of the Tetracyanoplatinate (II) Complex and its Oligomers, Using Electronic Structure Calculations, Emission Spectroscopy, and Solution X-ray Scattering* (Copenhagen, Denmark, 2010)
27. K.G. Dyall, K. Fægri Jr, Chem. Phys. Lett. **174**, 25 (1990)
28. K.G. Dyall, J. Chem. Phys. **100**, 2118 (1994)
29. K.G. Dyall, Theor. Chem. Acc. **112**, 403 (2004)
30. K.G. Dyall, Theor. Chem. Acc. **115**, 441 (2006)
31. T.H. Dunning Jr, J. Chem. Phys. **90**, 1007 (1989)
32. F. Weigend, R. Ahlrichs, Physical chemistry. Chem. Phys. **7**, 3297 (2005)
33. A. Schäfer, C. Huber, R. Ahlrichs, J. Chem. Phys. **100**, 5829 (1994)
34. D. Andrae, U. Haussermann, M. Dolg, H. Stoll, H. Preuss, Theor. Chim. Acta **77**, 123 (1990)
35. D. Figgen, K.A. Peterson, M. Dolg, H. Stoll, J. Chem. Phys. **130** (2009)
36. D. Paschoal, B.L. Marcial, J.F. Lopes, W.B. De Almeida, H.F. Dos Santos, J. Comput. Chem. **33**, 2292 (2012)
37. C. Heinemann, H. Schwarz, W. Koch, K.G. Dyall, J. Chem. Phys. **104**, 12 (1996)
38. C. Adamo, V.V. Barone, Theor. Chem. Acc. **105**, 169 (2000)
39. J.G. Snijders, J. Pyykkö, Chem. Phys. Lett. **75**, 5 (1980)
40. T. Ziegler, J.G. Snijders, E.J. Baerends, Chem. Phys. Lett. **75**, 1 (1980)
41. S.R. Bahn, K.W. Jacobsen, Comput. Sci. Eng. **4**, 55 (2002)
42. C. Peters, C.F. Eagen, Inorg. Chem. **15**, 4 (1976)

Chapter 4
X-Ray Scattering from Purely Classical MD

4.1 Probing Structural Change Using X-Ray Scattering

At the beginning of this project, the already established method within our group for fitting experimentally recorded XDS signals to molecular geometries was (very) simply put to (1) use DFT calculations to optimize geometries, (2) systematically modify important interatomic distances, and (3) calculate and compare the X-ray scattering from each modification step with the experimental data within a statistical χ^2-framework [1–6]. Scattering happens when an incident photon with wave vector $\mathbf{k_i}$ related to the radiation wavelength as $|\mathbf{k_i}| = \frac{2\pi}{\lambda}$ scatters off a particle, so the momentum transfer becomes $\mathbf{q} = \mathbf{k_s} - \mathbf{k_i}$, where $\mathbf{k_s}$ is the wave vector of the scattered photon. The isotropic scattering signal from a molecular structure is typically calculated via the Debye-equation [7].[1]

$$S(q) = \sum_i \sum_j f_i(q) f_j(q) \frac{\sin(qr_{ij})}{qr_{ij}} \tag{4.1}$$

with r_{ij} being the distance between atom i and j, and f being the atomic form factor. The next steps will progress as follows: We use this simple and established formulation on molecular structures obtained from MD simulations, and identify places where it can be improved. We then go back and take a closer look at the theory of X-ray scattering to make these improvements. Finally, we test the implementation of them.

As mentioned in the introduction, the experimental method employed by our group is of the X-ray 'pump-probe' kind: An ultra-short laser pulse—the 'pump' pulse, electronically excites the molecule at time $t = t_0$ and initiates the desired process. Then, at increasing times $t > t_0$, an X-ray probe pulse is scattered off the sample and recorded on a detector, giving the scattering signal $S_{on}(\mathbf{q})$, at time t.

[1] Scattering signals calculated with this formulation will from here on out be labeled 'Debye-scattering'.

A.O. Dohn, *Transient Changes in Molecular Geometries and How to Model Them*, Springer Theses, DOI 10.1007/978-3-319-18747-1_4

Since only a minor part of the molecules in the sample are excited by the pump pulse, the part of the $S_{on}(\mathbf{q})$-signal that expresses the structural change will be very small. Therefore, an 'off' signal of scattering from the unpumped sample, $S_{off}(\mathbf{q})$, is also recorded. The signal from everything that is unaffected by the pump, will then cancel out in the difference between these two signals, $\Delta S(\mathbf{q}) = S_{on}(\mathbf{q}) - S_{off}(q)$. Thus, on the probed time scale, if there is no change in the solvent part of the system under scrutiny, its contribution to the total scattering will be cancelled out in the difference scattering signal.

However, in a major part of the interesting processes of ultrafast molecular dynamics, the solvent *does play* a role, as discussed in the introduction. An example of solvent-interactions is the $[Fe(bpy)_3]^{2+}$complex, as described in Chap. 2, Sect. 2.3, where average, solvated solute-geometries simulated using an explicit solvation model within ab initio MD were readily obtainable, and the results of that study showing signs of both specific and non-specific solvation dynamic [8].

The objective here was to sample a sufficient amount of solvent configurations around the frozen solute- GS and ES-geometries using classical MD, and calculate the scattering signal from each step of each trajectory, to get an average solvent-contribution to the total difference scattering signal. Before delving into the results of using this strategy, and discussing its limits, we must briefly introduce the method of Molecular Dynamics simulations.

4.2 Molecular Dynamics

In its most basic form, MD simulations evaluate Newton's 2nd law of motion to propagate the particles in the system:

$$m_i \frac{d^2 \mathbf{r_i}(t)}{dt^2} = \mathbf{F}_i \tag{4.2}$$

where the forces $\boldsymbol{F}_i$ in this section are calculated from parametrised classical potentials, making up the so-called Molecular Mechanical force-fields.[2] The theory behind MD simulations is vast and well-documented [9–12], so instead of repeating it here, we limit this section to briefly describe the potentials used in a classical (MM) MD simulation.

The solvation-shell sampling was carried out in Desmond [13, 14] within the Schrödinger package [14], that uses the OPLS2005 [15–17] force-field for the solute. Like most force-fields, it divides up the energy E_{MM} in several terms of parametrised pair-wise additive potentials:

[2] In contrast to the main part of this work (part (iii) on Direct Dynamics), where the same differential equation is evaluated, the forces going into it are obtained from potentials calculated 'on-the-fly', using explicit calculations of the electronic structure of the relevant parts of the total system.

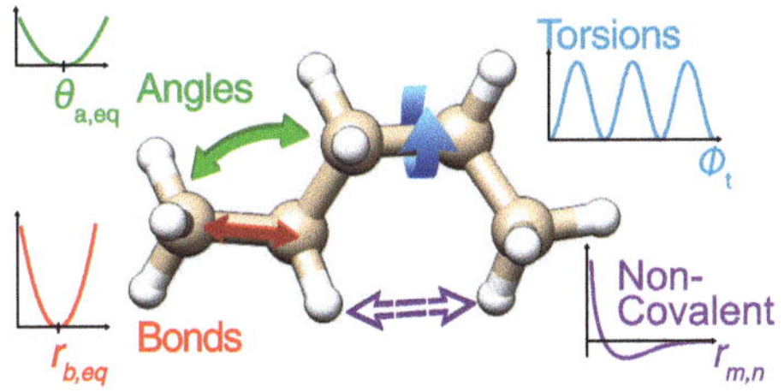

Fig. 4.1 An Illustration of the parametrised energy terms of Eq. 4.3. The indices a, b, t, m, and n run over the angles, bonds, torsions and atoms, respectively, and 'eq' represents the equilibrium value

$$\begin{aligned} E_{MM} &= \underbrace{E_{\text{bonds}} + E_{\text{angles}} + E_{\text{torsions}}}_{\text{Covalent}} + E_{\text{Non-Covalent}} \qquad (4.3) \\ &= \sum_{\text{b}}^{\text{bonds}} \frac{1}{2} K_b (r_b - r_{b,\text{eq}})^2 + \sum_{a}^{\text{angles}} K_a (\theta_a - \theta_{a,\text{eq}})^2 \\ &+ \sum_{t}^{\text{torsions}} \left(\sum_{j=1}^{3} \frac{V_{t,j}}{2} [(1 + \cos([j-1]\pi + j \cdot \phi_t)] \right) \\ &+ \sum_m \sum_{n \geq m} \left(\underbrace{\frac{q_m q_n}{r_{m,n}}}_{\text{Coloumb}} \underbrace{+ 4\epsilon_{m,n} \left[\left(\frac{\sigma_{m,n}}{r_{m,n}} \right)^{12} - \left(\frac{\sigma_{m,n}}{r_{m,n}} \right)^{6} \right]}_{\text{Lennard-Jones}} \right) f_{m,n} \end{aligned}$$

The terms are illustrated in Fig. 4.1. $f_{m,n}$ is a scaling factor which is 1.0 except for intramolecular '1, 4'-interactions between the end atoms involved in a torsional (dihedral) angle, where it is 0.5. The Lennard-Jones (LJ)-parameters ϵ and σ are tabulated for each element, and in OPLS they are combined using a simple geometric mean, e.g. $\epsilon_{m,n} = \sqrt{\epsilon_m \epsilon_n}$.[3]

However, since the parameters of this force-field are fitted to ground state ab initio calculations [16, 17], it would not make sense to use them in an attempt to predict excited state structures. Furthermore, since these force fields are often developed with the incentive of modelling large biomolecules, they cannot be assumed to accurately predict the interatomic distances around coordinated metals of varying exoticism. Therefore, as already mentioned, all the covalent energy terms were neglected by keeping the geometries of the complex fixed under the solvation-configuration sampling. Of course, this method of using a fixed solute will produce approximate solvent configurations, since the solute is not allowed to respond to collisions with solvent molecules.[4] Furthermore, and maybe more importantly, this way of approximat-

[3]More details about the non-covalent, or non-bonded interactions can be found in Sect. 5.5.1, Chap. 5, where they are tested for the QM/MM MD simulations.

[4]Compare throwing a ball into a brick wall to throwing a ball into a sheet of suspended cloth ....

ing the solvent shell (response) will only include steric effects from (changes in) the molecular geometry, since the force-field parameters (partial charges and the LJ parameters) will not reflect the difference in the electronic structure of any excited states.

4.3 X-Ray Scattering from MD Simulations of $[Fe(bpy)_3]^{2+}$

For the two production runs of each state of $[Fe(bpy)_3]^{2+}$, the mean solvated structures [8] were prepared in Desmond, and solvated using the TIP4P [18] water model, in cubes of 50 Å sides. The solvatisation was done using the built-in solvatisation function, which tiles up an already equilibrated pure solvent box over the final MD box, and deleting overlapping solvent molecules. Since the standard equilibration-scheme of the highly automated program does not take into account the necessity of having completely frozen solute structures through all the steps, a new version was made, based on the original, but kept the solute fixed during the entire equilibration procedure. A confirmation that the system still properly equilibrates is shown in Fig. 4.2.

The production runs were carried out in the NPT ensemble using the Martyna-Tobias-Klein barostat [19] and a Nose-Hover thermostat [20, 21], being the ensemble that most closely resemble experimental conditions, and the solvation shells were sampled every ps for 2 ns each, with a 2 fs integrator timestep. The histogram on the right side of Fig. 4.2 represents the distribution of instantaneous temperatures sampled every ps of the GS run. It can be shown [9], that for a canonical ensemble the variance σ_T^2 of the average temperature $\langle T \rangle^2$ should be:

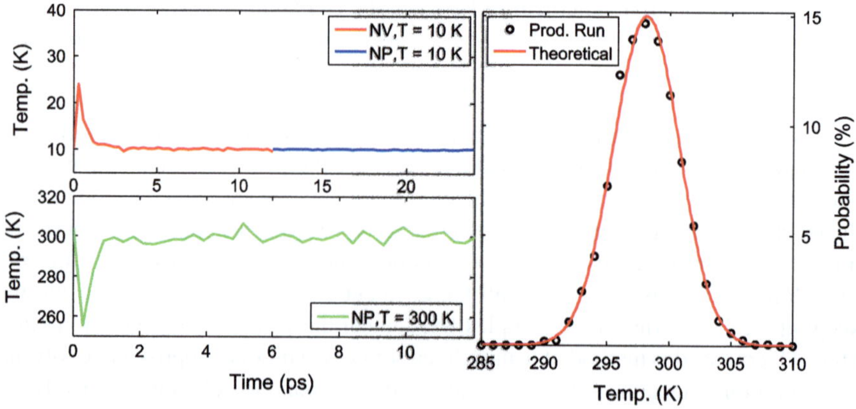

Fig. 4.2 Test of the modified equilibration scheme of Desmond, which keeps the solute frozen under all equilibration steps (the *red*, *blue*, and *green curves* on the *left* grafs), and also in the production run. The restraining is done by adding another potential $E_{res} = K_{res}(r - r_{res})^2$ to the total energy. For the equilibration, $K = 1000$ kcal/(mol · Å^2), and for the production, $K = 500$ kcal/(mol · Å^2)

$$\sigma_T^2 = \frac{2\langle T\rangle^2}{3N} \tag{4.4}$$

where $3N$ is the total number of degrees of freedom in the system, which amounts to 25197, according to the Desmond output, which automatically takes the restrained atoms into account. In the bulk limit of $3N \rightarrow \infty$, the variance becomes zero as expected for the thermodynamic temperature. The red Gaussian curve in the right part of Fig. 4.2 is made by simply plotting a normal distribution with the variance calculated according to Eq. 4.4 and the temperature of 298.15 K set in the thermostat, meaning that no fitting is involved, and the system exhibits the expected behaviour for a canonical ensemble of a finite system.

Figure 4.3 shows the Fe–N bond lengths of the two states, obtained by restraining the system, as described in Fig. 4.2. This method maintains the bond stretch induced by the electronic excitation, which is not describable with standard force fields. However, the thermal broadening represented by the variance of the bond distributions is completely controlled by the force constant chosen, and does as such not necessarily represent the actual physical broadening.

For each 35 ps of the two production runs, the X-ray scattering was calculated using the Debye equation. Figure 4.4 shows the results for each state. The low-q region on the plots the figure reveals spurious oscillations believed to arise from the finite box size. This dataset was handed over to the experimental part of our group for inclusion into the overall experimental analysis, where the non-physical oscillations were filtered out using Hann window [22], [I, VII].

The left part of Fig. 4.5 is taken from [I], and shows how the simulated signal *only* fits the data, if the contribution from the change in solvation-shell (green curve) is included. The right part of the figure shows the difference between simply calculating

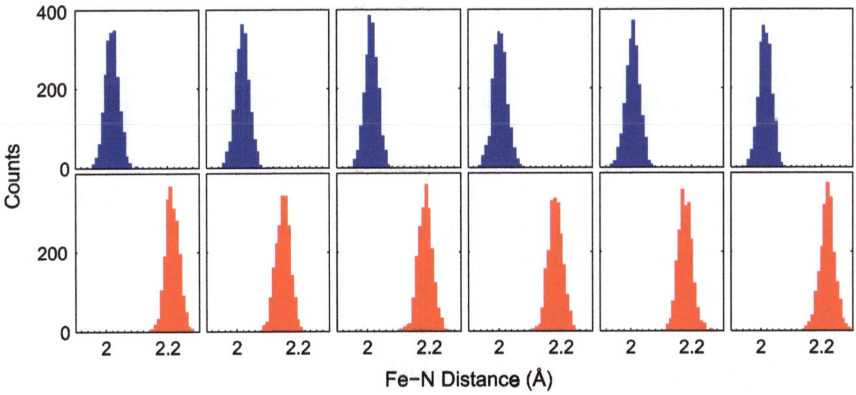

Fig. 4.3 Histograms of each of the 6 Fe–N bonds from the MD simulations using the 500 kcal/mol restraint on all $[Fe(bpy)_3]^{2+}$distances. *Blue* The GS geometry. *Red* ES geometry

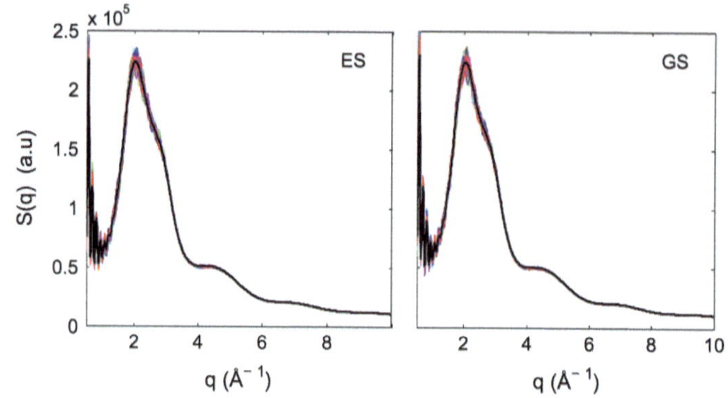

Fig. 4.4 Plots of the calculated scattering from snapshots of the MD trajectories. The *black curves* represent the averaged signal. Spurious oscillations with high-frequencies are observed at low q, originating from the unphysical truncation of the simulation box

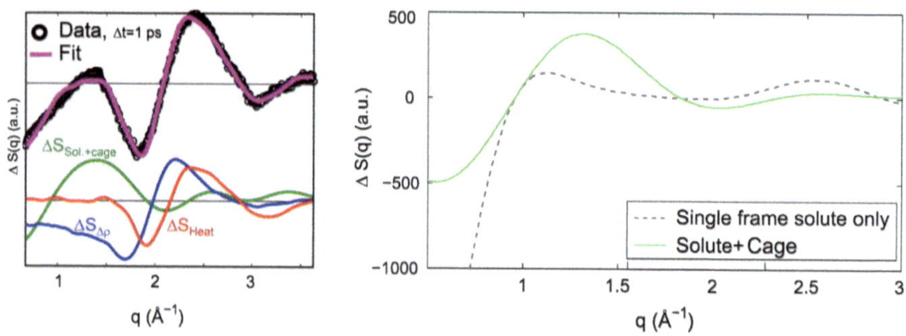

Fig. 4.5 *Left* The total XDS signal at $t = 1$ ps, with the fit carried out by the experimental part of our group [I]. *Right* The 'Solute+Cage' signal, i.e. the calculated average difference signal from the two trajectories containing ensembles of solvent configurations for each of the frozen structures (*green curve*). The *dashed grey curve* is calculated difference scattering from the naked solute structures obtained from [8], showing a dramatic change in the signal if the solvent is neglected

the signal from the structural change in the naked solute (dashed black line), and including the solvent shell response (green line). From this, it is evident that including solvent shell changes is required.

4.3.1 Improvements on the Calculation

The next steps in improving this method relies on approaching the following issues:

1. Using the Debye-formula for a system of 5000 particles requires computing the sine term in the rotational averaging 25 billion times pr time step.

2. The implementation of the Debye formula does not currently take into account the periodic boundary conditions of the MD simulations, but implementing this would increase the computational time further.
3. The truncation artifacts can have frequencies comparable to real, physical oscillations in the signal, making the heuristic Fourier filtering troublesome.

With regards to 1 and 2, methods akin to how long-range interactions in MD simulations are cut off could decrease the computational cost, alternative strategies centered around calculating $S(q)$ from the radial distribution function $g(r)$ is scattered throughout the literature [23–27]. Since $g(r)$ is ubiquitous in molecular simulations, methods and programs such as VMD that can calculate it efficiently are readily available [28, 29]. Therefore, we start the next section by going back to the theory of X-ray scattering, to derive an equation which is more suitable for use with MD simulations.

4.4 An Alternative to the Debye Formulation

$\Delta S(\boldsymbol{q}) = S_{on}(\boldsymbol{q}) - S_{off}(\boldsymbol{q})$, has been shown [30] to be a functional of the difference in nuclear density distribution of the pumped system, $\widetilde{\rho}_{on}(\boldsymbol{R}; t_p)$, and the unpumped, $\widetilde{\rho}_{off}(\boldsymbol{R}; t_p)$, and the molecular form factor $F(\boldsymbol{R}, \boldsymbol{q})$:

$$\Delta S(\boldsymbol{q}) = \int \left[\widetilde{\rho}_{on}(\boldsymbol{R}; t_p) - \widetilde{\rho}_{off}(\boldsymbol{R}; t_p)\right] |F(\boldsymbol{R}, \boldsymbol{q})|^2 \, d\boldsymbol{R} \tag{4.5}$$

Here, the nuclear density distribution $\widetilde{\rho}(\boldsymbol{R}; t_p)$ is a convolution of the instantaneous distribution of atomic positions at time t, $\rho(\boldsymbol{R}; t)$, with the intensity function $I(t)$ of the X-ray probe pulse:

$$\widetilde{\rho}(\boldsymbol{R}; t_p) = \int_0^\infty I(t)\rho(\boldsymbol{R}; \boldsymbol{t})dt \tag{4.6}$$

so in the approximation of an instantaneous X-ray pulse, given by the delta function $I(t) = \delta(t - t_p) \Rightarrow \widetilde{\rho}(\boldsymbol{R}; t_p) = \rho(\boldsymbol{R}; t_p)$.

The molecular form factor is in principle the expectation value of the scattering operator on the all electron wavefunction, which can be shown to give a Fourier transform of the electronic density [30]

$$|F(\boldsymbol{R}, \boldsymbol{q})|^2 = \left| \int \rho_e(\boldsymbol{r}; \boldsymbol{R}) e^{i\boldsymbol{q}\cdot\boldsymbol{r}} d\boldsymbol{r} \right|^2 \tag{4.7}$$

Almost always, the assumption is made that the scattering can be described as scattering from independent atoms, with spherical electronic densities. This is called the Independent Atom Model, and it effectively turns the molecular scattering factor into a sum of atomic form factors $f_i(q)$

$$F_{IAM}(\boldsymbol{R},\boldsymbol{q}) = \sum_j f_j(q)e^{i\boldsymbol{q}\cdot\boldsymbol{R}} \tag{4.8}$$

Even though this approximation ignores chemical bonding, it is in most cases very accurate for structural determinations [31], except maybe for liquid water [26], as shall become relevant later. While the electronic structure is directly available from DFT calculations, using it directly with an adequate numerical precision within a multidimensional structural-fitting strategy, as employed by the experimental section of our group [1], is unnecessarily cumbersome when working with systems where the Independent Atom Model (IAM) is sufficient.

Since Eq. 4.5 can easily be divided up in separate terms for $S_{on}(\mathbf{q})$ and $S_{off}(\mathbf{q})$, in the following we aim to derive an expression $S(\mathbf{q})$ usable in the context of molecular dynamics simulations of a solutes in liquid solvents. And since the quasi-structure inherent in liquids are often described via the Radial Distribution Function, $g(r)$[5] we aim to acquire an expression of the scattered X-ray intensity through $g(r)$. The definition of the Radial Distribution Function will also be a result of some of the steps in the derivation.

We start by assuming an instantaneous X-ray pulse at time t, and the IAM:

$$S(\mathbf{q}) = \int \rho(\mathbf{R})|F(\mathbf{R})|^2 d\mathbf{R} = \int \rho(\mathbf{R}) \sum_j f_j(\mathrm{q})e^{-i\mathbf{q}\cdot\mathbf{R_j}} \sum_k f_k(\mathrm{q})e^{i\mathbf{q}\cdot\mathbf{R_k}} d\mathbf{R} \tag{4.9}$$

In order to be able to do a separation of variables, we first separate the sums into sums of $j \neq k$ and $j = k$:

$$S(\mathbf{q}) = \int \rho(\mathbf{R}) \left(\sum_{j\neq k} f_j(\mathrm{q}) f_k(\mathrm{q})e^{-i\mathbf{q}\cdot(\mathbf{R_j}-\mathbf{R_k})} + \sum_{h=j=k} f_h(\mathrm{q})^2 \right) d\mathbf{R} \tag{4.10}$$

Since the last term is not dependent on $\mathbf{R}$, and the density is normalized, we are simply left with the sum, $\sum_{j=k} f_j^2(\boldsymbol{q})^2$.

We then focus on the first integral, and go to internal coordinates,[6] which means that the nuclear density is expressed as a function of all pairwise interatomic vectors $\mathbf{R_{j,k}} = \mathbf{R_j} - \mathbf{R_k}$, and the center of mass of the system, and separate the variables:

$$\rho(\mathbf{R}) \approx \rho(\boldsymbol{\Delta\mathcal{R}})\sigma(\mathbf{R_{cm}}) \tag{4.11}$$

where all the pairwise interatomic vectors $\boldsymbol{\Delta\mathcal{R}} = (\boldsymbol{\Delta}\mathbf{R_1}, \boldsymbol{\Delta}\mathbf{R_2}, \ldots)$ is a vector of $3\sum_j^{(N-1)} j$ coordinates, and $\boldsymbol{\Delta}\mathbf{R_i} = \mathbf{R_{k_i,j_i}}$, is the i'th vector between the k_i'th and j_i'th atomic pair. $\boldsymbol{\Delta\mathcal{R}}$ and $\mathbf{R_{cm}}$ thus describe all positions and angles between all the atoms in the system.

[5] Sometimes abbreviated as the Radial Distribution Function (RDF).

[6] This coordinate change is different from earlier methods [32], since the goal in this work is to end up with an expression of pair distributions.

When integrating over all space, there should also be equal probability for all possible centers of mass, since this essentially just is a translation of the entire sample.[7] In other words, σ is constant, and over all space, its integral must sum to 1. Now we are left with:

$$\sum_{j\neq k} f_j(\mathrm{q}) f_k(\mathrm{q}) \int \rho(\boldsymbol{\Delta}\mathcal{R}) e^{-i\mathbf{q}\cdot\mathbf{R_{j,k}}} d\boldsymbol{\Delta}\mathcal{R} \tag{4.12}$$

We now need to deal with the many-particle nucleic density $\rho(\boldsymbol{\Delta}\mathcal{R})$. If we assume that the total potential energy of the N-body system is pair-wise additive, we can write the distribution into the product for all pairwise distributions:

$$\sum_{j\neq k} f_j(q) f_k(q) \int_0^\infty \rho_1(\boldsymbol{\Delta}\mathbf{R_1})\rho_2(\boldsymbol{\Delta}\mathbf{R_2})\cdots e^{-i\mathbf{q}\cdot\boldsymbol{\Delta}\mathbf{R_{j,k}}} d\boldsymbol{\Delta}\mathbf{R_1} d\boldsymbol{\Delta}\mathbf{R_2}\cdots$$

Realising all other terms will integrate to 1, since each pair is dependent only on the $\mathbf{R}_{j,k}$ vector between atom j and k:

$$\sum_{j\neq k} f_j(q) f_k(q) \int_0^\infty \rho_{j,k}(\mathbf{R}_{j,k}) e^{-i\mathbf{q}\cdot\mathbf{R_{j,k}}} d\mathbf{R}_{j,k} \tag{4.13}$$

Which leaves us with a somewhat more approachable sum of integrals over each atomic pair and its corresponding probability density. Note that the approximation of pair-wise additivity has already been assumed within molecular dynamics simulations, since the forces propagating the system is calculated via the pair-wise additive potentials of the molecular force fields as seen in Sect. 4.2.

Now we make the isotropic assumption, which normally means that there is equal probability of finding the entire molecule in any orientation. This must mean that there is also equal probability of finding *each* $\mathbf{R}_{j,k}$ in any orientation, since a rotation of the molecule must mean a rotation of all of its pairwise vectors. For ultrafast studies, the isotropic assumption might not always hold, but since we are here trying to obtain an expression for the X-ray scattering formulated through the *radial* distribution function, this expression must be within the isotropic assumption. Options for including angular dependence in scattered intensity-expressions have been derived elsewhere [30, 32]. Evaluating the integral in the isotropic case will eventually lead to [30, 33]:

$$\sum_{j\neq k} f_j(q) f_k(q) \int_0^\infty \rho_{j,k}(r) \frac{\sin(qr)}{qr} dr \tag{4.14}$$

Remembering the second term where $j = k$ in 4.10 we get:

[7] From a theoretical viewpoint, and all other things being equal, the scattering is the same, regardless if it happens at SACLA in Japan or at LCLS in the States. The author is aware that this statement might be slightly provocative to some of our experimental collaborators.

$$S(q) = \sum_j f_j(q)^2 + \sum_{j \neq k} f_j(q) f_k(q) \int_0^\infty \rho_{j,k}(r) \frac{\sin(qr)}{qr} dr \tag{4.15}$$

We are still summing over all pairs in our system. Distribution functions often collect correlations of the same 'type' of atoms, which can be defined as the same element, e.g. as in the pairwise radial distribution function for the oxygen-oxygen correlation in water. Here, strictly speaking, one should define 'atom type' as 'atoms exhibiting the same scattering behaviour', which, within the IAM coincidentally would be the same as equating 'type' and element.[8] In other words, the minimum number of atom types in the system is the number of elements (ions), but it is possible to separate e.g. the solute and solvent terms by defining the atom types as 'elements belonging to the solute or the solvent'. Or more succinctly: You can have different atom types with the same atomic form factor, but you cannot have different form factors in the same atom type.

With this definition, we need redistribute the probability terms, such that all the N_l atoms j are of the same type as l, and all atoms k are the same type as m:

$$\rho_{l,m}(r) = \frac{1}{N_l(N_m - \delta_{l,m})} \sum_{\substack{j \in l \\ j \neq k}}^{N_l} \sum_{\substack{k \in m \\ k \neq j}}^{N_m} \rho_{j,k}(r) \tag{4.16}$$

Where $\delta_{l,m}$, the Kronecker delta, avoids the pairing of atoms with themselves. Note that l can be equal to m, or else we will miss the terms between different atoms of the same type, an error which seems to have crept into some formulations [23]. Since the form factors in Eq. 4.15 are not dependent on r, we can put them into the integral and rewrite the sum with the definition of atom types:

$$\sum_{j \neq k} f_j(q) f_k(q) \rho_{j,k}(r) = \sum_l \sum_m \sum_{\substack{j \in l \\ j \neq k}}^{N_l} \sum_{\substack{k \in m \\ k \neq j}}^{N_m} f_j(q) f_k(q) \rho_{j,k}(r)$$

where $f_j = f_l$ if $j \in l$ so:

$$= \sum_l \sum_m f_l(q) f_m(q) \sum_{\substack{j \in l \\ j \neq k}}^{N_l} \sum_{\substack{k \in m \\ k \neq j}}^{N_m} \rho_{j,k}(r)$$

and using the definition in Eq. 4.16:

$$= \sum_l \sum_m N_l(N_m - \delta_{l,m}) f_l(q) f_m(q) \rho_{l,m}(r) \tag{4.17}$$

which is then substituted into Eq. 4.15, where we also rewrote the sum of form factors squared for each atom, to the atom-type notation, which is evidently $N_l f_l(q)^2$:

[8] Or ion, since there is tabulated values for form factors for the same element with differing charges.

$$S(q) = \sum_l N_l f_l(q)^2 + \sum_{l,m} f_l(q) f_m(q) N_l (N_m - \delta_{l,m}) \int_0^\infty \rho_{l,m}(r) \frac{\sin(qr)}{qr} dr \tag{4.18}$$

To finally express the scattering in terms of the more used (pairwise) radial distribution function $g(r)$, we recall that it is defined via Eq. 4.19, which states that the number of particles (of a certain type B) dn at $r + dr$ (from the first type A) can be obtained by:

$$dn(r) = \sigma_0 g(r) 4\pi r^2 dr \tag{4.19}$$

where σ_0 is the isotropic number density, N/V. Therefore, we can similarly define

$$d\rho_{l,m} = \rho_0 g(r) 4\pi r^2 dr \tag{4.20}$$

where then ρ_0 is the isotropic probability density, $1/V$, and insert that into Eq. 4.18:

$$S(q) = \sum_l N_l f_l(q)^2 + \sum_{l,m} f_l(q) f_m(q) \frac{N_l (N_m - \delta_{l,m})}{V} 4\pi \int_0^R r^2 g_{l,m}(r) \frac{\sin(qr)}{qr} dr \tag{4.21}$$

The volume V normalises the radial distribution function, which means that the $g(r)$ peak amplitude contains information about the volume of the box used to calculate $g(r)$ in, so of course, the same volume must be used for both the $g(r)$ calculation and the following $S(q)$ calculation.

The integral in the above equation does not converge, but conveniently, one can rewrite the integral in Eq. 4.21 by adding and subtracting the distribution in the isotropic limit, $g_0(r)$ such that:

$$\int_0^\infty (g(r) - g_0) \frac{\sin(qr)}{qr} r^2 dr =$$
$$\int_0^\infty (g(r) - g_0) \frac{\sin(qr)}{qr} r^2 dr + \int_0^\infty g_0(r) \frac{\sin(qr)}{qr} r^2 dr$$

The last term has been argued to only contribute at $q \to 0$, which means we can exclude it [27, 34, 35], since this part of the q-range is covered in the experimental setup by the beamstop. Thus, we end up with:

$$S(q) = \sum_{l=m} N_l f_l(q)^2 + \sum_{l,m} f_l(q) f_m(q) \frac{N_l (N_m - \delta_{l,m})}{V} 4\pi \int_0^R r^2 [g_{l,m}(r) - g_{0,l,m}] \frac{\sin(qr)}{qr} dr \tag{4.22}$$

which can be implemented for numerical calculations of the scattering based on molecular simulations from which pairwise $g_{ij}(r)$ functions can be sampled. Often [23–26] (but not always [35, 36]), g_0 is simply written as 1, since $g(r)$ is normalised w.r.t the isotropic density. However, for the rewriting to work with respect to making the integral converge, the sampled $g(r)$'s must then also have converged to the isotropic limit of 1 at larger r's. This is rarely the case for solute-solute $g(r)$'s from MD simulations with a single solute in a large box of solvent molecules.

Equation 4.21 is actually a generalisation of the Debye-equation, which we will now show. The radial distribution function from a single pair of atoms i and j, i.e. the probability of finding atom j in the infinitesimal volume element $4\pi r^2 dr$ at distance r_{ij} from atom i must be a delta function[9]:

$$\frac{4\pi r^2 g_{ij}(r)dr}{V}dr = \delta(r - r_{ij})dr \tag{4.23}$$

which we can insert into Eq. 4.21:

$$\begin{aligned} S(q) &= \sum_{i=j} N_i f_i(q)^2 + \sum_{i,j} f_i(q) f_j(q) N_i (N_j - \delta_{i,j}) \int_0^\infty \delta(r - r_{ij}) \frac{\sin(qr)}{qr} dr \\ S(q) &= \sum_i \sum_j f_i(q) f_j(q) \frac{\sin(qr_{ij})}{qr_{ij}} \end{aligned} \tag{4.24}$$

Hence, of course, $N_i = N_j = 1$ for single atoms, and we have collected the two sums back into one term, since $\frac{\sin(qr)}{qr} \rightarrow 1$ for $r \rightarrow 0$.

Two issues arise when the numerical implementation is made:

1. The numerical representation of intramolecular $g_{ij}(r)$'s with delta-function-like characteristics depend on the numerical precision in $dr \approx \delta r$, the bin width for the sampled distributions. See Fig. 4.6 for a test of this.
2. The integral in Eq. 4.22 goes to infinity, but calculations are limited by the size of the simulation box.

The second issue introduces spurious truncation oscillations in the calculated scattering signal, since the integral is essentially the Fourier transformation of $g_{ij}(r)$. Many methods of varying heuristic nature has been applied to this problem, often for the reverse version of obtaining $g(r)$ from $S(q)$ [26, 37–42]. Some fit the tail of the data to an analytic function that can be continued to infinity [38], while others simply apply a dampening windowing function to the Fourier transformation [41, 42], and others again have developed more involved methods [26, 39, 40]. We have found it adequate so far to simply employ a window function $\frac{\sin(\pi r/R)}{\pi r/R}$ [42] in the transformation:

$$S(q) = \sum_{l=m} N_l f_l(q)^2 + \sum_{l,m} f_l(q) f_m(q) \frac{N_l(N_m - \delta_{l,m})}{V} 4\pi \int_0^R r^2 [g_{l,m}(r) - g_{0,l,m}] \frac{\sin(qr)}{qr} \frac{\sin(\frac{\pi r}{L})}{\frac{\pi r}{L}} dr \tag{4.25}$$

Some authors [42] choose R to be half the size of the simulation box, while others [41] provide no physical justification for their chosen value.

[9] Either atom j is there, or it is not, and the total possibility of finding the atom, i.e. integrating the distribution over all space, must be 1.

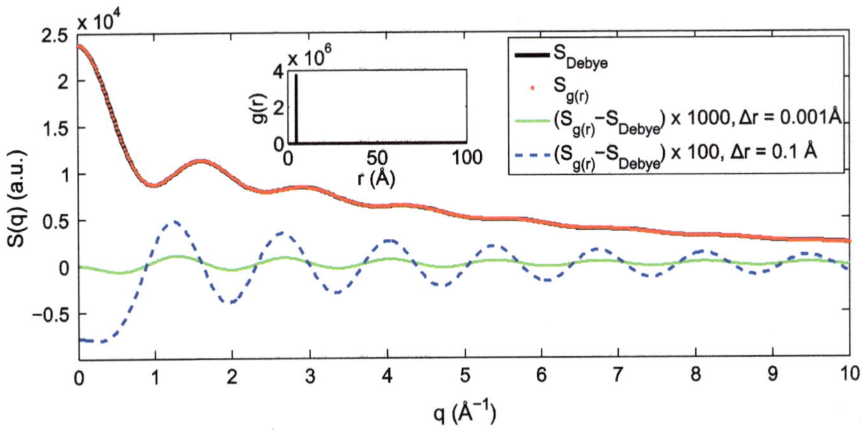

Fig. 4.6 Comparison of numerical implementations of calculated X-ray scattering, $S(q)$, implemented in Matlab. The scattered intensity goes to the number of electrons in the sample squared, when $q \rightarrow 0$. The estimated standard error of the calculation using the $g(r)$ with $\Delta r = 0.001$ Å is $2.12 \cdot 10^{-6}$. The *green curve* shows the residual. The *blue dashed curve* shows the residual when using $\Delta r = 0.1$ Å, giving a standard error of $1.37 \cdot 10^{-4}$. The *inset* shows the $g(r_{\Delta=0.001})$-function used in the calculation

4.4.1 Testing the Implementation

Figure 4.6 shows a comparison of the numerical implementation of Eq. 4.25 made in the Matlab program, with a numerical implementation of the Debye-formula (Eq. 4.24), as previously made and used in our group. The system is very simple, simply two Ir atoms at distance $r = 4.63$ Å. For $S(q)$ calculated via Eq. 4.25, $g(r)$ was numerically calculated with $\delta r = 0.01$ Å, going from 0 to 100 Å, in a square box of 100 Å side lengths (see inset on Fig. 4.6), thus numerically approximating the delta-function. Based on the observation that for $q \rightarrow 0$, the scattering goes to $f(q \rightarrow 0)^2$. Remembering that, in principle, $f(q) = \int \rho(r)e^{iqr}dr$, where $\rho(r)$ is the electronic density, then $S(q \rightarrow 0) \rightarrow n^2$ where n is the number of electrons in the system.[10] Thus, it is confirmed that the two implementations produce the same scattering in the limit of a single atomic pair, as is seen in Fig. 4.6.

The next step involves calculating the scattering of neat water, as shown in Fig. 4.7. Here, an experimentally resolved $g_{OO}(r)$-curve [43, 44] has been digitized and used in our numerical implementation. The two other pairwise radial distribution functions of water was unfortunately not resolved during that experiment, so instead we have used curves simulated using the TIP4P-eW force field, which is made by parts of the same group [45] who authored the experimental publication. The results are displayed with the dashed grey line on the left graph in the figure. Since the O–O correlations contain the most electrons, this term should make up the major part of the scattered intensity. Another experiment from the literature [46], provided a

[10]The unscaled Debye-scattering in Fig. 4.6 at $q = 0.02$ Å is $2.37 \cdot 10^4 = (77 \cdot 2)^2$.

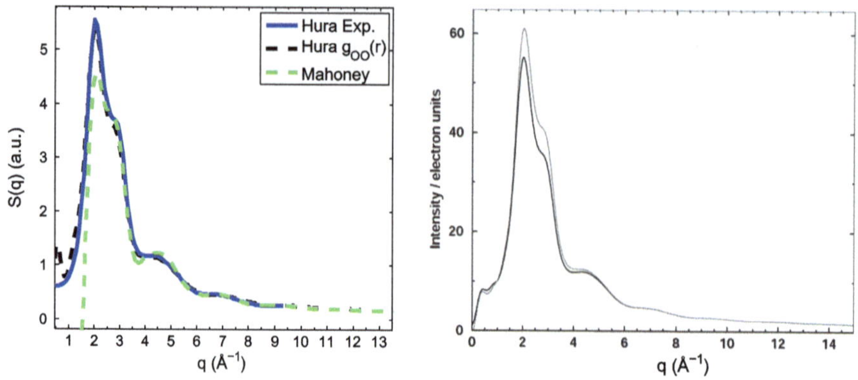

Fig. 4.7 *Left* Scattered X-ray intensity from neat water, experimentally obtained by Hura et al. (*blue line*) [43, 44]. From that experiment Hura et al. resolved the O–O pairwise RDF, which was then used in our numerical implementation, to give the *dashed black curve*. The *dashed green line* represents scattering calculated via pairwise RDF$_s$ obtained by Mahoney and Jorgensen [46]. This data lacks the delta-functions from intramolecular O–H and H–H correlations, which causes the errors in low-q. *Right* A figure from Hura et al. [44], comparing calculated scattering within the IAM (*black line*) and with a Modified Atomic Form Factor, that takes chemical bonding into account

full set of experimentally resolved pairwise radial distribution functions, which is also included in the figure. For larger q, there is overall a good agreement between the experimental scattering obtained by Hura and coworkers [43, 44] (blue line), and the corresponding simulated scattering. The small differences between the blue and dashed black curves are remarkably similar to the differences observed by Hura et al. (Fig. 4.7, right), comparing simulated scattering within the IAM approximation (black line) with a Modified Atomic Form Factor [44], that takes redistribution of charge due to chemical bonding into account. In a pump-probe experiment, however, the scattering contribution from the (unchanged) bulk solvent away from the solute is cancelled out, since its contribution to $S_{\mathrm{on}}(q)$ and $S_{\mathrm{off}}(q)$ is identical.

This concludes the benchmark of the new numerical implementation. This method of calculating scattered X-ray intensity via pairwise radial distribution functions will hopefully be a valuable addition to the toolbox of our group, since it can provide a calculation of the experimental signal from solvation shell changes, as well as change in the solvent itself, provided molecular simulations of the process of interest is carried out.

4.5 Returning to $[Fe(bpy)_3]^{2+}$ and Concluding

In the case for $[Fe(bpy)_3]^{2+}$from Sect. 4.3, we compare the scattering signal calculated with new method to the one used in the article in Fig. 4.8. In the previous simulations, the simulation box volume was only recorded for the final step of the simulation. The new method for calculating the signal relies on the simulation box volume to correctly scale the radial distribution functions, so new MD runs were made in the NVT ensemble. Also, for the remainder of the projects presented, ensembles with fixed volumes have been used.

The black curve represents the total difference signal calculated with a dampening of 20 Å. Considering the severe truncation effects at low q for the Debye-scattering (Fig. 4.4), it is possible that the filtering also removed some of the real signal, seeing as the amplitude at larger q is larger in the $g(r)$-version in Fig. 4.8. However, the two calculated signals are qualitatively very similar, and both reproduce the experimental low-q increase in scattering.

At this point it is instructive to further analyse exactly which solvent shell changes we are simulating, since the MM MD approximation used to obtain a better sampling of the RDF_s do not account for the changes in the electronic structure undergone by the molecule in the excited state. Figure 4.9 compares the RDF_s from the ab initio MD study [8] to results from the NVT simulations used in the scattering comparison. For $g_{Fe–O}(r)$, the main loss of the solvent shell peak around 5 Å is reproduced by the MM model, but the ES $g_{Fe–O}(r)$ increases in amplitude at shorter distances, compared to the GS, and since the minimum after the first peak disappears in the excited state, the MM model produces a more diffuse first solvent shell around the excited state structure. The ground state solvent shell minimum distance is roughly 6.3 Å in the ab initio model, and around 5.8 Å in the MM model.

The MM $g_{Fe–H}(r)$ again shows qualitative similarity, but a more well-defined solvent shell for the MM case. All in all, at least parts of the non-specific solvent shell changes are recovered in the MM approximation, as described in the above.

The dashed lines in Fig. 4.9 represent the cumulative coordination numbers, $c_n(r)$. The number $n(r)$ of atoms (of the correct type) in a spherical shell with borders from r_1 to r_2 is simply obtained by integrating both sides of Eq. 4.19:

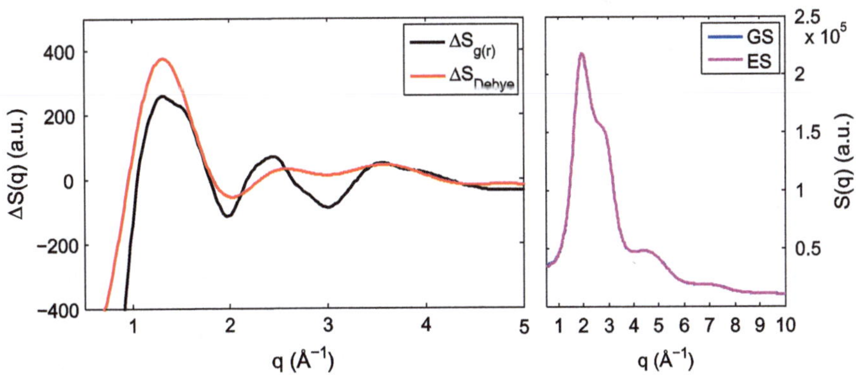

Fig. 4.8 *Left* Difference scattering signal calculated from MD simulations redone in the NVT ensemble, compared to Debye-scattering from the first MD runs. *Right* The full signals

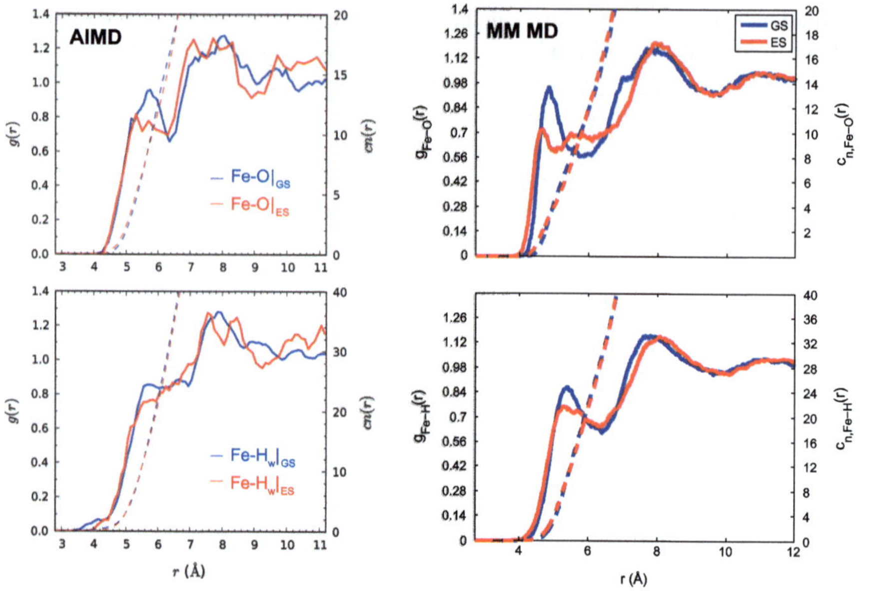

Fig. 4.9 *Left* The RDFs obtained via an ab initio MD method [8]. *Right* RDFs from the MM MD simulations based on the molecular structures obtained from [8]. While the two models produce qualitatively similar results, the specific solvation dynamics are not reproduced, since there is effectively no change in coordination number in the first solvation shell, in the MM model

$$n(r) = \int_{r_1}^{r^2} \sigma_0 g(r) 4\pi r^2 dr \tag{4.26}$$

and the coordination number in solvents are often defined by setting r_1 to 0 and r_2 to the first minimum in $g(r)$.

The change in Fe–O coordination number from GS to ES for the MM model is 0.1, meaning that within the approximation of only describing changes in the molecular structure, and not in the electronic, the specific solvent dynamics are not accounted for. This must again mean that the specific expulsion of the water molecules from the first solvation shell upon excitation, observed both in the computer [8] and in the laboratory [VII, I], must be heavily influenced by changes in the *electronic* structure of the excitation, and not only by the changes in molecular geometry. Obviously, this separation between the electronic and 'geometric' structure (i.e. the nuclear coordinates) is an artificial one, brought upon the system by the need for a statistically satisfying sampling of the solvent configurations for the scattering calculation.

As already alluded to, the chosen strategy for improving on the models for the transient changes exhibited in molecules like $[Fe(bpy)_3]^{2+}$ has been to employ the Direct Dynamics method of the following sections of this work. Like the similar ab initio MD method to which the MM MD results where compared, this method will explicitly include changes in the electronic structure of the solute, affected by the solvent charges.

References

1. K. Haldrup, M. Nielsen, *Measuring and Understanding Ultrafast Phenomena Using X-rays*, pp. 91–113. NATO Science for Peace and Security Series A: Chemistry and Biology (Springer Science+Business Media B.V., 2014). ISBN 978-94-017-8549-5
2. K. Haldrup, T. Harlang, M. Christensen, A. Dohn, T.B. van Driel, K.S. Kjær, N. Harrit, J. Vibenholt, L. Guerin, M. Wulff et al., Inorg. Chem. **50**, 9329 (2011)
3. K. Haldrup, M. Christensen, M.M. Nielsen, Acta Crystallographica A **66**, 261 (2010)
4. A. Dohn, N. Harrit, *Structural Investigations of the Tetracyanoplatinate(II) Complex and its Oligomers, Using Electronic Structure Calculations, Emission Spectroscopy, and Solution X-ray Scattering* (Denmark, Copenhagen, 2010)
5. T.B. van Driel, N. Harrit, *Time-Resolved Structural Analysis of $Rh_2(dimen)_4^{2+}$ in solution* (Denmark, Copenhagen, 2010)
6. T. Harlang, N. Harrit, *Spectroscopic and X-ray Structural Investigations of $[Rh_2(dimen)_4]^{2+}$ and $[Ir_2(dimen)_4]^{2+}$ on Pico- and Femtosecond Timescales* (Denmark, Copenhagen, 2010)
7. J. Als-Nielsen, D. McMorrow, *Elements of Modern X-ray Physics*, 1st edn. (Wiley, New York, 2001). ISBN 0471498580
8. L.M.L. Daku, A. Hauser, J. Phys. Chem. Lett. **1**, 1830 (2010)
9. D. Frenkel, B. Smit, *Understanding Molecular Simulation* (Academic Press, San Diego, 2002). ISBN 0-12-267351-4
10. M. Allen, D.J. Tildesley, *Computer Simulation of Liquids* (Oxford University Press, Oxford, 1989). ISBN 0-19-855645-4
11. H.J.C. Berendsen, *Simulating the Physical World* (Cambridge Press, Cambridge, 2007). ISBN 9780521835275
12. J.H. Jensen, *Molecular Modeling Basics* (CRC Press, Boca Raton, 2010)
13. D. Shivakuamr, J. Williams, Y. Wu, W. Damm, J. Shelley, W. Sherman, J. Chem. Theory Comput. **6**, 1509 (2010)
14. *Suite 2012: Desmond Molecular Dynamics System*, version 3.1 (2012), D.E. Shaw Research, New York, NY, 2012; *Maestro-Desmond Interoperability Tools*, version 3.1, Schrödinger, New York, NY
15. J.L. Banks, H.S. Cao, A.E. Cho, W. Damm, R. Farid, A.K. Felts, T.A. Halgren, D.T. Mainz, J.R. Maple, R. Murphy et al., J. Comput. Chem. **26**, 1752 (2005)
16. W.L. Jorgensen, J. Tirado-Rives, J. Am. Chem. Soc. **110**, 1657 (1988)
17. W.L. Jorgensen, D.S. Maxwell, J. Tirado-Rives, J. Am. Chem. Soc. **118**, 11225 (1996)
18. W.L. Jorgensen, J. Am. Chem. Soc. **103**, 335 (1981)
19. G.J. Martyna, D.J. Tobias, M.L. Klein, J. Chem. Phys. **101**, 4177 (1994)
20. S. Nosé, J. Chem. Phys. **81**, 511 (1984)
21. W.G. Hoover, Phys. Rev. A **31**, 1695 (1985)
22. A.V. Oppenheim, R.W. Schafer, J.R. Buck, *Discrete-Time Signal Processing* (Prentice Hall, Upper Saddle River, 1999), p. 468. ISBN 0-13-754920-2
23. H. Ihee, M. Wulff, J. Kim, S. Adachi, Int. Rev. Phys. Chem. **29**, 453 (2010)
24. T.K. Kim, M. Lorenc, J.H. Lee, M.L. Russo, J. Kim, M. Cammarata, Q. Kong, S. Noel, A. Plech, M. Wulff et al., Proc. Nat. Acad. Sci. **103**, 9410 (2006)
25. J.M. Sorenson, G. Hura, R.M. Glaeser, T. Head-Gordon, J. Chem. Phys. **113**, 9149 (2000)
26. T. Head-Gordon, G. Hura, Chem. Rev. **102**, 2651 (2002)
27. N.S. Gingrich, Rev. Mod. Phys. **15**, 90 (1943)
28. W. Humphrey, A. Dalke, K. Schulten, J. Mol. Graph. **14**, 33 (1996)
29. B.G. Levine, J.E. Stone, A. Kohlmeyer, J. Comput. Phys. **230**, 3556 (2011)

30. K.B. Møller, N.E. Henriksen, Struct. Bond. **142**, 185 (2012)
31. P. Coppens, Ann. Rev. Phys. Chem. **43**, 663 (1992)
32. U. Lorenz, K.B. Møller, N.E. Henriksen. New J. Phys. **12**, 113022 (2010)
33. J. Taylor, *Scattering Theory: the Quantum Theory on Nonrelativistic Collisions* (Wiley, New York, 1972)
34. N.S. Gingrich, B.E. Warren, Phys. Rev. **46**, 248 (1934)
35. H.P. Klug, L.E. Alexander, *X-ray Diffraction Procedures: For Polycrystalline and Amorphous Materials* (Wiley, New York, 1974). ISBN 978-0-471-49369-3
36. R.F. Kruh, Chem. Rev. **62**, 319 (1962)
37. B.E. Warren, *X-Ray Diffraction* (Dover Publications, New York, 1990)
38. A.H. Narten, C.G. Venkatesh, S.A. Rice, J. Chem. Phys. **64**, 1106 (1976)
39. P.F. Peterson, E.S. Božin, T. Proffen, S.J.L. Billinge, Appl. Crystallogr. **36**, 53 (2003)
40. J. Köfinger, G. Hummer, Phys. Rev. E **87**, 052712 (2013)
41. J.H. Lee, K.H. Kim, T.K. Kim, Y. Lee, H. Ihee, J. Chem. Phys. **125**, 174504 (2006)
42. G. Gutiérrez, B. Johansson, Phys. Rev. B **65**, 104202 (2002)
43. G. Hura, J.M. Sorenson, R.M. Glaeser, T. Head-Gordon, J. Chem. Phys. **113**, 9140 (2000)
44. G.L. Hura, D. Russo, M. Glaeser, T. Head-Gordon, M. Krack, M. Parrinello, Phys. Chem. Chem. Phys. **5**, 1981 (2003)
45. H.W. Horn, W.C. Swope, J.W. Pitera, J.D. Madura, T.J. Dick, G.L. Hura, T. Head-Gordon, J. Chem. Phys. **120**, 9665 (2004)
46. M.W. Mahoney, W.L. Jorgensen, J. Chem. Phys. **112**, 8910 (2000)

Part III
Direct Dynamics

Chapter 5
Background

This work is mostly focused on application, which is why this chapter is intended as a guide for future users, with only a brief theoretical outline of the methods employed in this part of the project. The theory behind the GPAW simulations in this chapter is explained in detail elsewhere [1–3], and a thorough overview of the QM/MM implementation can be found in [4]. As GPAW is a DFT-based code, the basics from Chap. 3 are also applicable here.

5.1 The Grid-Based Projector Augmented Wave Method

Most of the computational chemistry programs, like the ones used in Chap. 3, follow along the same strategy, and solve the Kohn-Sham equation of a system of non-interacting electrons using a basis set made from a linear combination of atomic orbitals, often represented by a(nother) linear combination of Gaussian functions. GPAW takes a somewhat different approach in dealing with the problem of calculating electronic structure of a many-body system. The difference is twofold:

1. GPAW uses a soft valence pseudo-wave function-description while retaining the cusps and sharp features of electrons in the core region. The two regions are smoothly connected by the PAW transformation [5, 6].
2. The pseudo-wave functions are evaluated directly, by discretizing the Kohn-Sham equations, and solving them numerically using the FD method. The uniform, real-space grids can be surprisingly coarse due to the smoothness of the pseudo-wave function outside the core region (see Fig. 5.1). This makes the calculations faster, and more parallelisable between many CPU cores.

5.1.1 PAW

Exactly how the PAW method matches up the two regions is beyond the scope of this work, but more info can be found in [1–3, 5, 6]. However, the result is that

A.O. Dohn, *Transient Changes in Molecular Geometries and How to Model Them*, Springer Theses, DOI 10.1007/978-3-319-18747-1_5

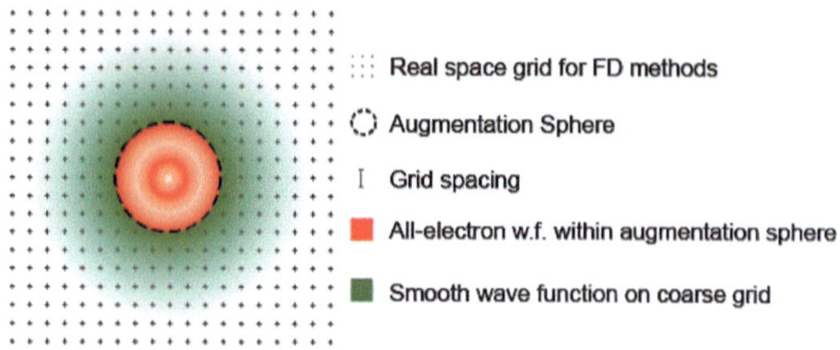

Fig. 5.1 The basics of GPAW. The pseudo-wave function is described numerically on an uniform grid, and the all electron core states within the augmentation region are described on a finer, radial grid, and can be evaluated independently of the neighbouring environment, before the actual calculation, and saved in a GPAW 'setup'. The PAW method makes sure that the two regions are smoothly matched

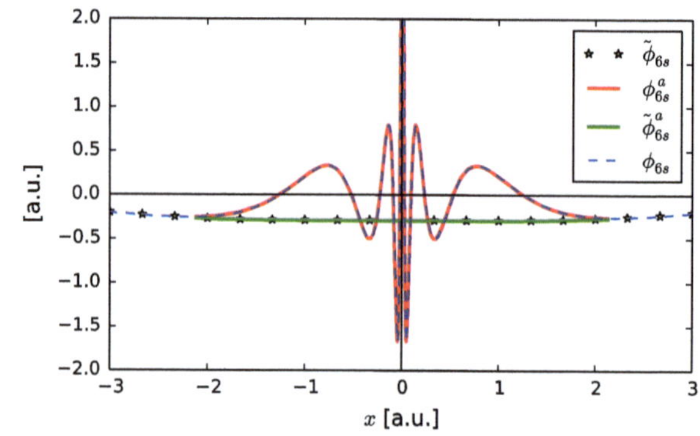

Fig. 5.2 An example of how the 6s orbital of Pt is computed within the PAW formulation. The plot shows the real values of the various wave functions as a function of the distance x from origo. The a superscript demarcates the wave functions within the atom-specific 'augmentation sphere', and the tilde-marked are the pseudo-wave functions. From [7]

atomic (core region) variables can be calculated independently of the neighbouring environment, and prior to calculations of entire systems. This saves a considerable amount computational expense. These so-called 'PAW atomic setups' are functional-dependent, since the values like contain core densities, atomic kinetic energy contributions etc., which depend on which functional is used to calculate them.

Figure 5.2 shows the resulting 6s orbital of Pt, made as illustrated in Fig. 5.1:

$$\phi_{6s} = \tilde{\phi}_{6s} + \phi^a_{6s} - \tilde{\phi}^a_{6s} \tag{5.1}$$

$\tilde{\phi}$ is the all-electron partial pseudo-wave, the ϕ^a is the all-electron wave within the 'augmentation sphere', and $\tilde{\phi}^a$ is the part of the pseudo-wave within the augmentation sphere that needs to be subtracted to end up with the overall behaviour of ϕ_{6s}.

For spatially large systems, requiring large QM cells, GPAW has an LCAO-mode [8] option, akin to more traditional computational chemistry codes. This will make GPAW use a basis of LCAO, which speeds up calculations. However, the grid-based methods are still being used for the density and the potential.

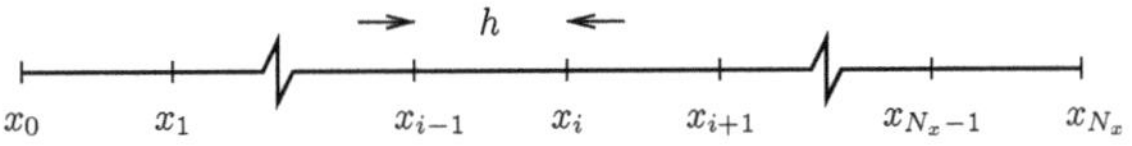

Fig. 5.3 A one-dimensional grid used as an example for the 3D mesh used in GPAW

5.1.2 Finite Difference and Real Space Grids

The FD method is well known and heavily used within the engineering community as well as in physics, and since the PAW method allows for such coarse grids to be employed in the valence regions that the FD method is viable, GPAW can take advantage of the already well-established, efficiency-boosting grid-methods [9–11]. Here, we only briefly line up the basic FD principle, which is very much akin to the discretization of time that takes place in the dynamic integrators for MD.

As an example of where the method can be used, the Kohn-Sham kinetic energy operator in Eq. 5.19 involves the Laplacian. However, first, let us just look at the general example of how to numerically approximate a first order derivative, in one dimension on a grid like the one in Fig. 5.3, before dealing with the second order derivative. If we make a Taylor expansion of the function we want to numerically evaluate:

$$T(x_i + h) = T(x_i) + h \frac{\partial T(x)}{\partial x}\bigg|_{x_i} + \frac{h^2}{2} \frac{\partial^2 T(x)}{\partial x^2}\bigg|_{x_i} + \cdots \tag{5.2}$$

Ignoring everything more that first order terms and isolating the first order derivative gives the forward difference formula:

$$\frac{\partial T(x)}{\partial x}\bigg|_{x_i} \approx \frac{T(x_i + h) - T(x_i)}{h} \approx \frac{T(x_{i+1}) - T(x_i)}{h} \tag{5.3}$$

Using the notation $T(x_i + h) = T(x_{i+1})$. Obviously the derivative can also be approximated from functional values in the point before x_i:

$$\frac{\partial T(x)}{\partial x}\bigg|_{x_i} \approx \frac{T(x_i) - T(x_{i-1})}{h} \tag{5.4}$$

Finally, the numerical precision can be increased by using a point on each side of x_i to evaluate the value of the derivative at x_i

$$T(x_i + h) - T(x_i - h) \approx T(x_i) + h \, \frac{\partial T(x)}{\partial x}\bigg|_{x_i} - T(X_i) + h \, \frac{\partial T(x)}{\partial x}\bigg|_{x_i}$$

$$\Rightarrow$$

$$\frac{\partial T(x)}{\partial x}\bigg|_{x_i} \approx \frac{T(x_{i+1}) - T(x_{i-1})}{2h} \tag{5.5}$$

which is recognised as a basic definition of a derivative, if $h \to 0$.

Second order derivatives can be approximated by including one more term of the Taylor expansions of $T(x_i + h)$ and $T(x_i - h)$, and adding them up:

$$T(x_i + h) + T(x_i - h) \approx$$

$$T(x_i) + h \left.\frac{\partial T(x)}{\partial x}\right|_{x_i} + \frac{h^2}{2}\left.\frac{\partial^2 T(x)}{\partial x^2}\right|_{x_i} + T(x_i) - h \left.\frac{\partial T(x)}{\partial x}\right|_{x_i} + \frac{h^2}{2}\left.\frac{\partial^2 T(x)}{\partial x^2}\right|_{x_i}$$

$$\Rightarrow$$

$$\left.\frac{\partial^2 T(x)}{\partial x^2}\right|_{x_i} \approx \frac{T(x_{i+1}) + T(x_{i-1}) - 2T(x_i)}{h^2} \tag{5.6}$$

This can readily be extended to more dimensions, and to an arbitrary number of neighbouring grid points used:

$$\left.\frac{\partial^k T}{\partial x^k}\right|_{x_i} \approx \frac{1}{h^k} \sum_{j \in N_j} \alpha T_j \tag{5.7}$$

where the coefficients α_j are determined by the Taylor expansion, and N_j is a list of grid points used in the approximation, termed the *stencil*. E.g. the 2 order derivative above uses a 3-point stencil of $T(x_{i-1})$, $T(x_i)$ and $T(x_{i+1})$.

Using a real space grid method has the added advantage that, when treating the electrostatic effect of classical point charges on the electronic density, the density is already represented on a grid.

5.2 The Combined Quantum Mechanical/Molecular Mechanical Method

The advent of combining quantum- and molecular mechanical approaches began with the seminal work by Warshel and Levitt [12] that recognized the need for a QM treatment of some, *but not all* regions of large, biological systems, such as reaction centers in enzymes. Traditional classical force field methods, as described in Chap. 4, are computationally inexpensive, but cannot make or break bonds, describe excited states without laborious re-parametrization of the particle interactions, and are overall less general than QM methods. On the other hand, the increased computational cost of QM methods make them infeasible for very large systems evolving in time.

Therefore, Warshel and Levitt combined the QM and MM descriptions, choosing a chemically active region to be quantum mechanically described, while the rest of the system (coupled to the QM part) is described classically. This idea was quickly expanded upon, and numerous variations have been developed and implemented [13–16]. Ultimately, Warshel and Levitt, were together with Martin Karplus awarded the Nobel prize in chemistry in 2013 [17].

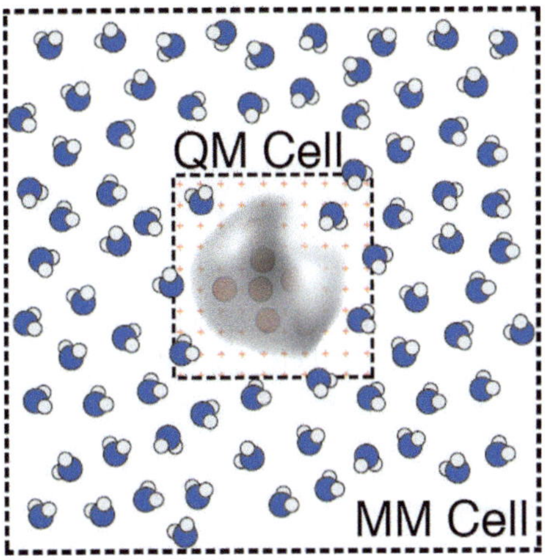

Fig. 5.4 A Schematic representation of the main idea behind the hybrid QM/MM strategy. The total system to be simulated is divided up in two subsystems: *1* The electrostatically embedded QM subsystem, where the electronic structure is calculated using DFT. *2* The MM subsystem which is described using classical force-field methods. The two subsystems are interacting according to the description in the text

Here, we aim to present a general introduction to how interactions between the two sub-systems can be handled, since a more technically detailed description of the used implementation has already been presented [4]. The entire system is sketched in Fig. 5.4. The total energy of the system cannot simply be described as the sum of the energies of the QM and MM subsystems, since the two regions are of course interacting. In this implementation, the total energy of the system is based on an additive scheme:

$$E_{\text{tot}} = E_{\text{QM}} + E_{\text{QM/MM}} + E_{\text{MM}} \tag{5.8}$$

The interaction term between the two subsystems $E_{\text{EQ/MM}}$ is comprised of the:

1. Electrostatic interactions between the QM electronic density and the MM point charges, E_1^{EL}.
2. The electrostatic interactions between the QM nuclei and MM point charges, E_2^{EL}.
3. van der Walls (vdW) interactions between the QM and MM particles, E_{LJ}.

So, for the electronic density $\rho(\mathbf{r})$, the MM point charges q_m with positions $\mathbf{R}_m$, and the nucleic charges in the QM subsystem Z_n with positions $\mathbf{R}_n$[1]:

$$\begin{aligned} E_{\text{QM/MM}} &= E_1^{\text{EL}} + E_2^{\text{EL}} + E_{\text{LJ}} \\ &= \sum_m^{N_{\text{MM}}} q_m \int \frac{\rho(\mathbf{r})}{|\mathbf{r} - \mathbf{R}_m|} d\mathbf{r} + \sum_m^{N_{\text{MM}}} \sum_n^{N_{\text{QM}}} \frac{q_m Z_n}{|\mathbf{R}_n - \mathbf{R}_m|} + E_{\text{LJ}} \end{aligned} \tag{5.9}$$

[1]Atomic units are still used here, since carrying around $1/(4\pi\epsilon_0)$ is cumbersome ....

where the umbrella-termed vdW interactions from the many-body effects Pauli repulsion and dispersion are simply modelled with a generic LJ 12-6 potential, just like the LJ term in Eq. 4.3 of Sect. 4.2:

$$E_{\text{LJ}} = \sum_{m}^{N_{\text{MM}}} \sum_{n}^{N_{\text{QM}}} 4\epsilon_{mn} \left[\left(\frac{\sigma_{mn}}{|\mathbf{R}_n - \mathbf{R}_m|} \right)^{12} - \left(\frac{\sigma_{mn}}{|\mathbf{R}_n - \mathbf{R}_m|} \right)^{6} \right] \tag{5.10}$$

The electronic density is self-consistently converged under the influence of an external potential $V'_{ext} = V_{ext} + V_{\rho-MM}$, where V_{ext} is the original potential, and $V_{\rho-MM}$ describes the influence on the MM point charges on the electronic density:

$$V_{\rho-\text{MM}}(\mathbf{r}, \mathbf{R_m}) = \frac{\partial E_{\text{QM/MM}}[\mathbf{R_n}, \mathbf{R_m}]}{\partial \rho(\mathbf{r})} = \sum_{m}^{N_{\text{MM}}} \frac{q_i}{|\mathbf{r} - \mathbf{R_m}|} \tag{5.11}$$

where $\mathbf{r}$ is on the grid. Divergences caused by point charges very close to grid points (termed *the charge spillout effect* [18]) is treated by smoothening the potential in Eq. 5.11, using an established method [18], and tested by E. Jónsson, to give consistent potential minima of a variety of QM/MM interactions [4], when using a smoothening factor of 0.2 Å. The smoothed potential is slightly more involved, but here we aim to explain the basic methodology using the simpler, unsmoothed potential.

The long range electrostatics are treated according to the MIC [19] (see Fig. 5.5), but since the computational bottleneck here is the SCF cycles, a point charge simply interacts electrostatically with all other point charges within its minimum image, as described in the right side of Fig. 5.5. If *very* large MM cells are to be used in the future, it would be advantageous to implement a long-range cutoff.

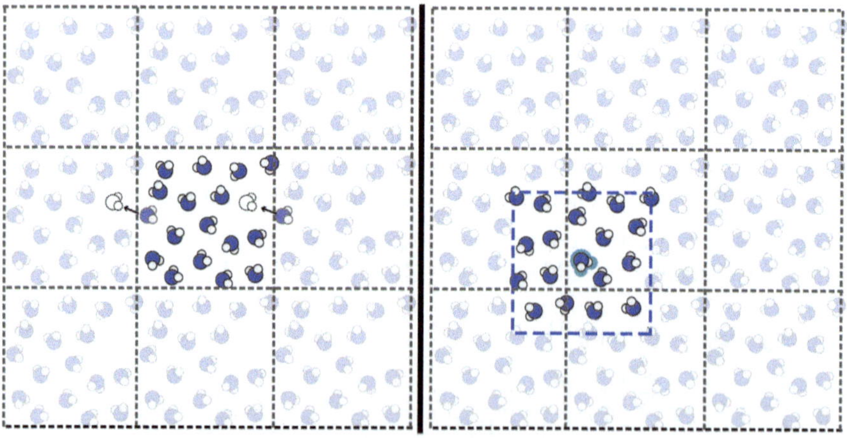

Fig. 5.5 *Left* Schematic of the periodic boundary conditions. *Right* Schematic of how long-range electrostatics are cut off in the MIC. The highlighted molecule is electrostatically affected by the fully drawn molecules, comprising a minimum image with the original box dimensions

5.3 Born-Oppenheimer Molecular Dynamics

The review of Senn et al. from 2009 states that '*first-principles QM/MM MD remains computationally demanding even by today's standards*' [14]. This is because each time step in the dynamics requires that the electronic density is re-optimized within the SCF scheme. The efficiency of GPAW is vital in this regard. This stepwise re-calculation of the electronic density for each set of nucleic coordinate at time t_i is very much akin to the Born-Oppenheimer approximation, which is already used in the basic DFT theory to simply treat the nuclei via the stationary external potential they create. As such, the resulting QM/MM MD motion is completely adiabatic. Too large time steps will worsen the chances of converging the electronic density.

The nuclei in the QM subsystem are still treated classically within this method, and therefore also propagated according to the classical forces they experience. So, in the QM subsystem, for nucleus n:

$$\mathbf{F_n} = \mathbf{F}_n^{\rho} + \mathbf{F}_n^{n} + \mathbf{F}_{n-\text{MM}}^{\text{QM/MM}} = -\frac{\partial E_{\text{QM}}[\mathbf{R_n}]}{\partial \mathbf{R_n}} - \frac{\partial E_{\text{QM/MM}}[\mathbf{R_n}, \mathbf{R_m}]}{\partial \mathbf{R_n}} \tag{5.12}$$

where it is recalled that the first, purely QM energy term is still affected by the MM subsystem through the extra potential term (Eq. 5.11), and is handled in the original GPAW code, which readily provides $\mathbf{F}_n^{\rho} + \mathbf{F}_n^{n}$ for a QM nucleus n with charge Z_n. The second term of Eq. 5.12 is then

$$-\frac{\partial E_{\text{QM/MM}}[\mathbf{R_n}, \mathbf{R_m}]}{\partial \mathbf{R_n}} = Z_n \sum_{m}^{N_{\text{MM}}} \frac{q_m}{|\mathbf{R_n} - \mathbf{R_m}|^2} - \frac{\partial E_{\text{LJ}}(\mathbf{R_n}, \mathbf{R_m})}{\partial \mathbf{R_n}} \tag{5.13}$$

which accounts for the forces $\mathbf{F}_{n-\text{MM}}^{\text{QM/MM}} = \mathbf{F}_{n-\text{MM}}^{coloumb} + \mathbf{F}_{n-\text{MM}}^{\text{LJ}}$ from the classical point charges on the nucleus n of the quantum system.

The forces on the classical point charges follow along analogously, but the derivatives are with respect to the position of the point charges. So for point charge m:

$$\mathbf{F_m} = \mathbf{F}_m^{\rho} + \mathbf{F}_{\text{MM}-n}^{\text{QM/MM}} + \mathbf{F}_m^{m} \tag{5.14}$$

The pure MM term is readily handled within ASE, so we focus simply on the new additions made in this implementation: $\mathbf{F}_m^{\rho} + \mathbf{F}_{\text{MM}-n}^{\text{QM/MM}}$:

$$\mathbf{F}_m^{\rho} + \mathbf{F}_{\text{MM}-n}^{\text{QM/MM}} = - q_m \int \frac{\rho(\mathbf{r})}{|\mathbf{r} - \mathbf{R_m}|^2} d\mathbf{r} - q_m \sum_{n}^{N_{\text{QM}}} \frac{Z_n}{|\mathbf{R_n} - \mathbf{R_m}|^2} - \frac{\partial E_{\text{LJ}}(\mathbf{R_n}, \mathbf{R_m})}{\partial \mathbf{R_m}} \tag{5.15}$$

The numerical integral represents the most computationally expensive step of the QM/MM interfacing, since it runs over all the grid points for each MM point charge.

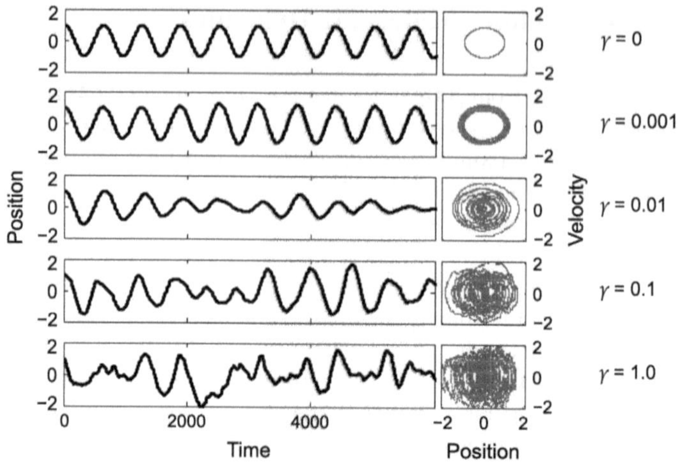

Fig. 5.6 Plots of the effect of increasing the friction term γ in Eq. 5.16, illustrated by a simple numerical implementation of the equation, using a harmonic potential. The same integrators as in ASE is used. $\gamma = [0; 0.001; 0.01; 0.1; 0.5]$, from *top* to *bottom*

The purely classical force term $\mathbf{F}_m^m$ is handled as in the traditional ASE code, and the LJ forces from the last terms of both Eqs. 5.13 and 5.15 are simply obtained by taking the derivative of Eq. 5.10.

The particles of the system are then propagated according to the forces evaluated as described above.[2]

5.3.1 Molecular Dynamics in ASE

The Langevin-type dynamics module of ASE has been modified by Jónsson [4] to work within the QM/MM framework, and an option of constraining interatomic distances, using RATTLE-type [20] constraints has been added. The Langevin equation of motion solved numerically in ASE has the following form:

$$m_i \frac{d^2\mathbf{r_i}(t)}{dt^2} = \mathbf{F}_i - \gamma m_i \frac{d\mathbf{r_i}(t)}{dt} + \sqrt{2k_B T \gamma m_i}\mathbf{R}(t) \tag{5.16}$$

$\mathbf{F}_i$ is the total force on the particle, obtained as described in the previous section, T is the desired temperature, and $\mathbf{R}(t)$ is a stochastic force term. This term has its physical origin in Brownian motion, which describes the diffusive behaviour of a particle experiencing random collisions with its neighbours. γ is called the friction coefficient; increasing this value effectively decreases the strength of the inertial (deterministic) forces and increases the stochastic forces. The effect of the friction is illustrated in Fig. 5.6, where the motion of a single particle in a 1D harmonic

[2]Remembering the added technical complexity of short-range smoothening.

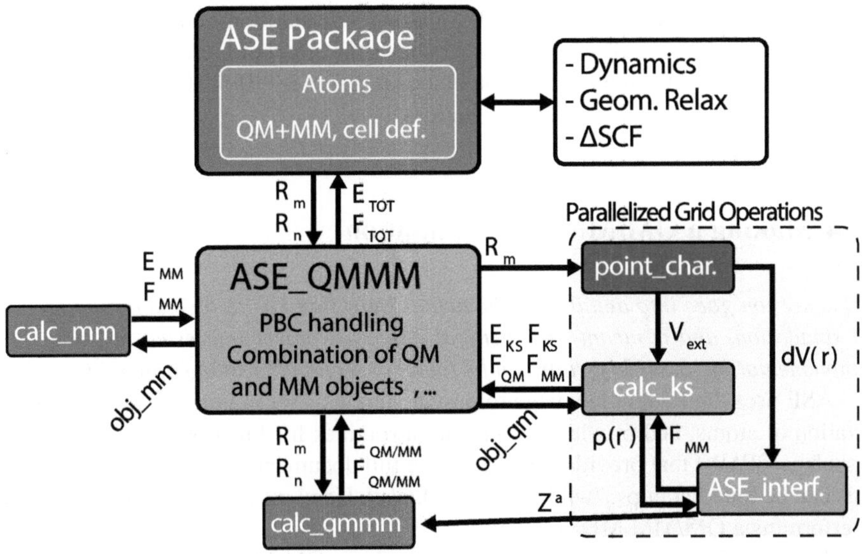

Fig. 5.7 Schematic of the QM/MM code; how it operates within ASE, and together with GPAW. The main interfacing takes place in ase_qmmm, where the two ASE objects are created from the subsystems, the PBC are handled in the presence of the QM cell, and calls to the calculators are defined. calc_mm is the calculator for the MM subsystem, and calc_ks for the QM subsystem. So far, the interfacing works together with the ASE dynamics module, but could be readily expanded to work with other tasks, such as geometry relaxation, ΔSCF methods for approximation of excited states [22, 23], etc. Adapted from [4]

potential is simulated using the same type of numerical integrators as in ASE, i.e. the Velocity-Verlet integrators [21]

$$x(t + \Delta t) = x(t) + v(t)\Delta t + \frac{1}{2}\frac{F(x)}{m}\Delta t^2$$
$$v(t + \Delta t) = v(t) + \frac{1}{2}\left[\frac{F(x(t + \Delta t)) + F(x(t))}{m}\right]\Delta t$$

Langevin dynamics is of the canonical (NVT) ensemble, and the Langevin thermostat can be kept on in production runs where only information about thermodynamic ensemble averages are of interest. However, running Langevin-type simulations to describe the accurate atomic movements on the microscopic scale would be pointless, since the temperature is kept constant by applying random forces to the system. Luckily, the microcanonical (NVE) ensemble is easily recovered, simply by setting γ to 0 and obtaining Newton's equation of motion:

$$m_i \frac{d^2\mathbf{r_i}(t)}{dt^2} = \mathbf{F}_i \tag{5.17}$$

which produces the expected sinusoidal motion for the harmonic oscillator, as seen in the top plot of Fig. 5.6. The figure further shows the effect of increasing the friction.

A schematic of the implementation is shown in Fig. 5.7. How the implementation is used is described in the following sections.

5.4 Making a QM/MM MD Simulation

This section goes into detail with the actual steps that has to be taken for running a simulation, and is meant as a general guide for others interested in using this implementation. Sample scripts can be found in section B.1 in the appendix.

ASE uses the Python language to provide modules for manipulation and visualization of atoms, and provides optimization routines for functions, or calculations (such as GPAW) that provides energies. The full documentation, and helpful tutorials can be found at https://wiki.fysik.dtu.dk/ase/. Here, we line up the main steps of performing a QM/MM MD simulation.

There are four major steps involved in running a QM/MM MD simulation using this (or any other) implementation:

1. Prepare a MM solvent box
2. Analyse optimal QM parameters
3. Embed QM subsystem in MM subsystem, equilibrate
4. Simulate production run.

5.4.1 Preparing Solvent Boxes

This is done by preparing a script that spans out a grid in a box of the desired size, and then puts a solvent molecule on each grid point. This initial configuration is obviously very far from the thermodynamic equilibrium of the system. This means that the initial forces will be very large, which will then require very small time steps of the numerical integrator to initially propagate the system, in order to avoid a build-up of numerical errors that will eventually crash the simulation. In the early days of MD simulations, it made sense to spend time 'pre-randomizing' the initial configuration, to cut down on the computational expense of this initial 'thermalization'. With the computing power presently available, purely classical MD simulations are cheap and fast, so spending more 'human'-time preparing the system is much more expensive than spending the 'silicon'-time thermalizing it from a worse starting point. The lesson is well remembered, however, when switching to the hybrid QM/MM MD, where converging the electronic density is a serious bottleneck, which is why embedding the QM subsystem in the MM box should be done at the latest possible point in the procedure.

5.4.2 Analysing Optimal QM Parameters

As with any computational study, it is essential to ascertain a robust description of the system under investigation. In the GPAW framework, this means considering choice of functional, grid spacing, basis set size (if LCAO-mode is employed), and the QM cell size. The choice of all these parameters should then also be judged in relation to the computational feasibility of MD simulations. This is essential for obtaining a model within feasible limits of computational costs. While GPAW *is* fast, QM/MM MD is by no means 'cheap'. As an example, the simulations in the following section sum up to over 30 CPU years of computation-time. An example of how these choices can be considered in detail is given for $[Ir_2(dimen)_4]^{2+}$ in Sect. 6.1.

5.4.3 Embedding QM Subsystem in the MM Box

This step requires reading in the thermalized MM solvent box, the (previously structurally relaxed) system chosen for the QM description, and adding them up, deleting overlapping solvent molecules. Another equilibration run is needed to let the two subsystems respond to the sudden presence of each other.

The QM/MM implementation is written such that, in ASE terminology, the total atoms object should start with the QM subsystem, and then contain MM subsystem. Lastly, the atomic indexes of all the atoms in the total system should *not* change/be changed between steps, and the sequence of atoms in the solvent part should also be kept (i.e. for ACN: MeCN, MeCN,...).

5.4.4 The Production Run

When the previous steps have been taken care of, the following choices can be addressed:

- How large time steps can be taken without the drift (i.e. the numerical error of the discrete integration) becoming a problem?
- How much can the number of needed SCF cycles be reduced by tuning the density mixing scheme, that mixes in electronic density from previous steps in the SCF-cycle, and/or the convergence criteria?
- How small can the QM cell be without truncating the wave functions?

The output files from the equilibration and thermalization runs can be helpful to analyse for drift and optimization of GPAW parameters to reduce the number of needed SCF steps before convergence is reached.

A template for an input script for a QM/MM MD simulation can be found in Section B.1 Appendix B.

5.5 Implementing an Acetonitrile Force Field

Since the predominantly used solvent for $[Ir_2(dimen)_4]^{2+}$ is ACN, a classical, rigid, 3-site interaction potential was adapted from Guardia et al. [24], where the non-bonded (LJ) parameters of the methyl group is collected into a single site. The parameters employed are shown in Table 5.1, and were derived from ab initio Hartree-Fock calculations on a gas phase ACN dimer with a 6-31G** basis.

5.5.1 *Benchmarking the Acetonitrile Liquid Pseudo-Structure*

To test whether the implementation of the force field could successfully reproduce the liquid pseudostructure of ACN, a simulation box of $28 \times 28.5 \times 31.5$ Å was filled with 290 ACN molecules and thermalized. A MM production run of 0.5 ns was simulated. The simulation was used to confirm the expected thermodynamic behaviour of the model, (Fig. 5.8, right), and used to sample the RDFS. The translational speeds of the center of mass of each molecule is compared to the Maxwell-Boltzmann distribution:

$$f(v) = 4\pi v^2 \sqrt{\left(\frac{m}{2\pi k_B T}\right)^3} e^{-\frac{mv^2}{2k_B T}} \tag{5.18}$$

where m is the total mass of an ACN molecule, v is the speed, and k_B is the Boltzmann constant. The result is shown in Fig. 5.8, revealing a good correspondence between the simulation and the expected distribution. The instantaneous kinetic temperature is calculated via $\langle E_{\text{kin}} \rangle = \frac{3}{2} k_b T \Rightarrow T = \frac{m}{3k_b} \langle v^2 \rangle = 299.8$ K, only 0.2 K from the temperature set in the thermostat.

In order to analyse the QM/MM interfacing, RDF between a single QM ACN and the remaining MM ACN were produced, and compared to RDF obtained only using MM MD.

Parallel QM/MM runs with a single QM ACN were branched off the production MM trajectory, interspaced by 500 fs, to avoid any strong correlation between the new trajectories (see the velocity autocorrelation function inset on Fig. 5.9). The Langevin-type thermostat was used in both the MM and QM/MM runs, with a friction

Table 5.1 The non-bonded Molecular Mechanics parameters, and intermolecular fixed distances, from Guardia et al. [24] ϵ is the potential well-depth and σ the interaction distance

Site	ϵ (kJmol^{-1})	σ (Å)	q (e)
Me	0.7825	3.775	0.206
C	0.5440	3.650	0.247
N	0.6276	3.200	−0.453
$r_{\text{MeC}} = 1.458$ Å		$r_{\text{CN}} = 1.157$ Å	

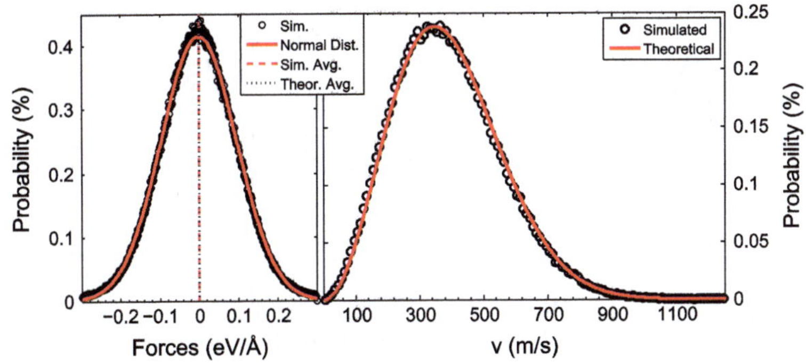

Fig. 5.8 *Left* Histogram of all the forces in the QM/MM simulations of a single QM ACN in an MM ACN box. The QM-to-MM interfacing is working properly if there is no average force on the entire system. The avg. force from the simulation is 4.41×10^{-8} eV/Å. *Right* A histogram of the translational speeds of the center of mass of each ACN molecule in the purely classical NVT simulation, plotted with the theoretical Maxwell-Boltzmann distribution

coefficient of 0.05 and 2 fs timesteps. The hydrogens on the QM were constrained using the RATTLE scheme [20]. Each QM/MM trajectory was thermalized again, before sampling the RDFs. A total of 0.5 and 0.25 ns of dynamics were sampled for the MM and QM/MM systems, respectively.

The first test of a successful QM/MM interfacing can be found in the left part of Fig. 5.8, which shows a histogram of all the forces on all the molecules of the total of 0.25 ns of QM/MM trajectory sampled. The average force on the entire box should be zero, as it should of course not 'move' with time. The average force is 4.41×10^{-8} eV/Å, and the normal distribution of forces in the Langevin-type ensemble that result in the Maxwell-Boltzmann distribution of velocities is properly reproduced.

Figure 5.9 shows an overall good agreement between the various descriptions. The literature RDFS [24] have been produced in the NPT ensemble, which might be the cause of some of the small differences. Another dissimilarity between the simulations is how the LJ parameters are combined for each pair of site types, for the LJ energy term in Eq. 5.10. Guardia et al. uses the Lorentz-Berthelot rules [26, 27]:

$$\sigma_{ij} = \frac{\sigma_{ii} + \sigma_{jj}}{2} \qquad \text{and} \qquad \epsilon_{ij} = \sqrt{\epsilon_{ii}\epsilon_{jj}} \tag{5.19}$$

—or simply an arithmetic and geometric mean for σ and ϵ, respectively. The combination-rules employed here are of the Waldman-Hagler type, which should provide a better overall description of interactions between a broader range of systems [28, 29]:

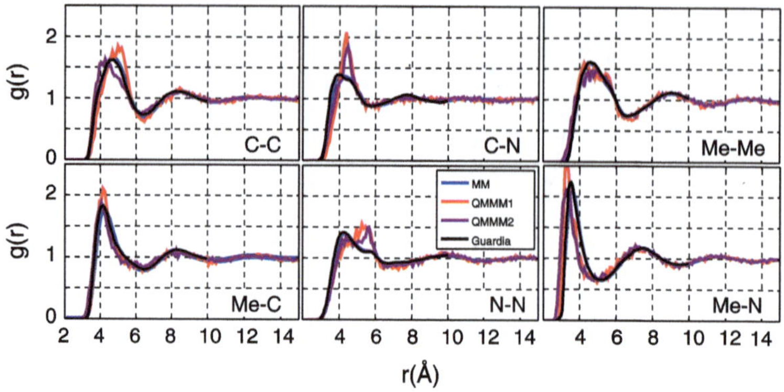

Fig. 5.9 Radial distribution functions (RDF) of the 6 intermolecular distances of ACN. The RDF are calculated every 500th fs in trajectories of 0.5 and 0.25 ns for the MM and QM/MM systems, respectively. The QM/MM systems are comprised of a single QM ACN in a bath of MM ACN. The QM/MM1 systems are made using methyl-group vdW parameters for the middle carbon [24], while QM/MM2 uses aliphatic parameters [25]. There is overall a good agreement between the various descriptions

$$\sigma_{ij} = \left[\frac{\sigma_{ii}^6 + \sigma_{jj}^6}{2}\right]^{1/6} \quad \text{and} \quad \epsilon_{ij} = \frac{2\sigma_{ii}^3\sigma_{jj}^3}{\sigma_{ii}^6 + \sigma_{jj}^6}\sqrt{\epsilon_{ii}\epsilon_{jj}} \tag{5.20}$$

The difference between using these two combination schemes are shown in Fig. 5.10, where the LJ potential has been plotted for the interaction between the methyl group and the nitrogen. The difference in the resulting potential from the combined LJ parameters is mainly noticeable in the repulsive region, with Methyl-N distances which would be energetically highly unfavourable in the first place. At k_BT for 300 K, the difference r between the two combined potentials is 4 pm.

A last consideration is that the electronic structure description used to create the force field is not the same as the one used in the QM/MM simulations, and the force field is not optimized to work in a QM/MM framework. This will be discussed further in the following section.

5.5.2 Acetonitrile Dimer Binding Curves

The difference in the QM and MM description is explained through a study of the binding energy curves for two ACN geometries, as depicted in Fig. 5.11. The cyclic version has the lowest energy [30–33], while the colinear version must be oriented along the dipole moment.

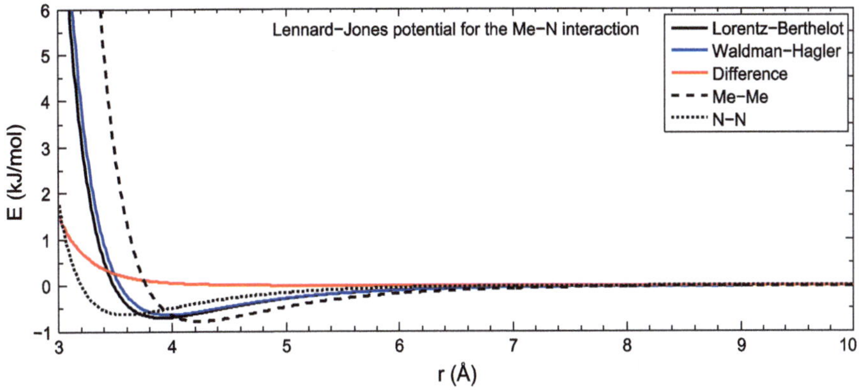

Fig. 5.10 Lennard-Jones potential for the nonbonded Me-N interactions in acetonitrile, combined using the Lorentz-Berthelot- and Waldman-Hagler rules, respectively. The difference is largest in the very repulsive region

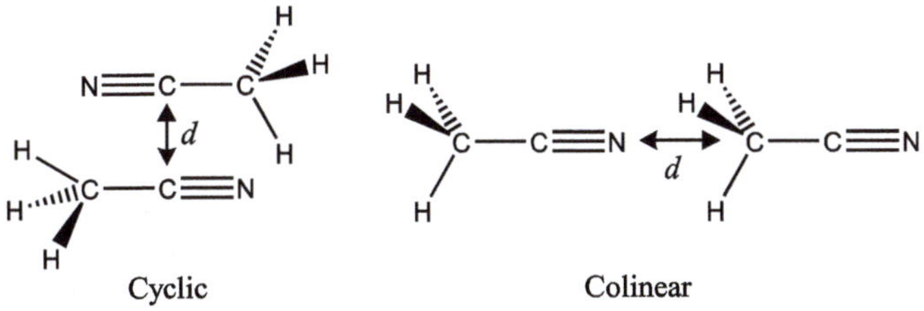

Fig. 5.11 Two known ACN dimers, where the cyclic version is believed to have the lowest energy [30–33]. The dimer binding energy is obtained by subtracting 2 times the potential energy of the single molecule from the total potential energy of the dimer, at increasing distances d

The binding energy U_{bind} is simply:

$$U_{\text{bind}} = U_{\text{dimer}}(d) - 2U_{\text{single,QM}} \tag{5.21}$$

where $2U_{\text{single,QM}}$ is the potential energy of a single ACN, and $U_{\text{dimer}}(d)$, is the dimer potential energy at intermolecular distances d.

For the purely QM calculation, both ACN are described using DFT with the PBE functional, at a grid spacing of 0.18 Å. In the LCAO-mode calculations, the tzp basis is used. The QM cell was set to 20 Å.

For the QM/MM calculations, one ACN is QM, while the other is MM. Here, $U_{\text{bind}} = U_{\text{dimer}} - U_{\text{single,QM}}$, since the MM ACN has no intramolecular energy terms—all bonds and angles are simply fixed.

For the same reason, the purely MM/MM binding energy is simply the dimer potential energy. The results are shown in Fig. 5.12. The QM/MM minimum for the linear dimer is slightly lower in energy, and larger in r, which would result in a slight overbinding. This is in agreement with the QM/MM $g_{C-N}(r)$ peak in Fig. 5.9 being

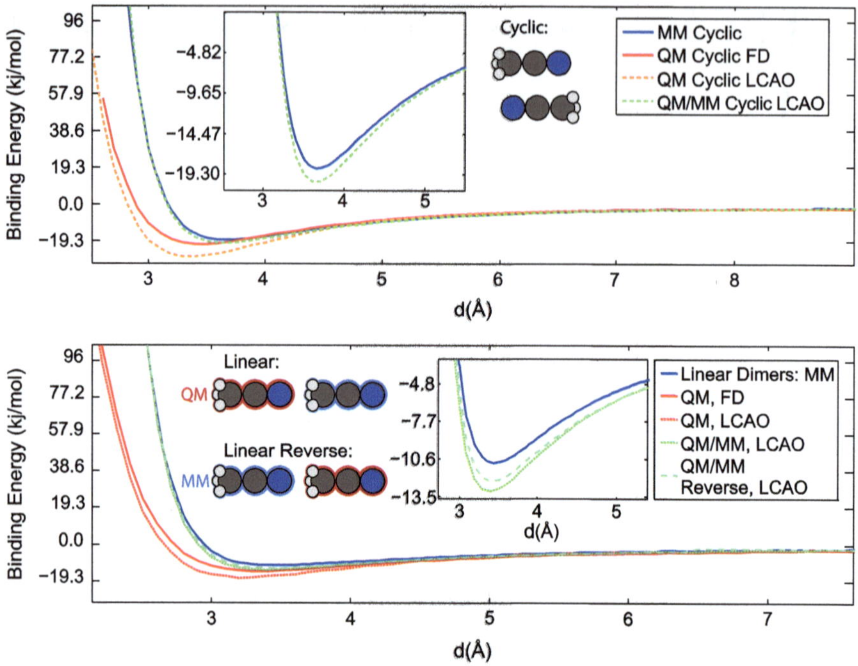

Fig. 5.12 Binding energy curves of ACN. *Top* The cyclic dimer. *Bottom* The linear dimer. Since there are two ways of combining the QM and the MM description, both the 'linear' and the 'linear reverse' dimers have been defined as in the graph, and examined. The insets depict are zoom-ins on the largest differences between the MM/MM and QM/MM curves. The 'FD' curves are made in the pure grid-based finite difference mode, whereas the LCAO-labeled curves uses basis sets from linear combinations of atomic orbitals for the initial wave function guesses

too well-defined, and peaking at a slightly longer distance than the MM equivalent. Overall, the slight overbinding of both QM/MM dimers is generally reflected in the less diffuse QM/MM radial distribution functions.

The purely quantum mechanical description of the dimer overbinds the dimer much more, a well-known effect from another polar (and popular) solvent: Water [34–37]. As such, it can be assumed that simply extending the QM subsystem to encompass some (more, or all) of the solvent does not necessarily increase the precision of the description, unless long-range corrected functionals are employed. On the contrary, the solvation shell description might actually suffer from simply including solvent molecules in a QM subsystem using the PBE functional.

It must be mentioned that the electronic density of a dimer in vacuum cannot be assumed to be identical to that of a dimer surrounded by other solvent molecules. Thus, one should be careful of postulating too much about the full QM solvent pseudo-structure from the energetics of a dimer, but since water and acetonitrile share similar physical properties (hydrogen bonding, high polarity), the conjecture could be made, that a fully QM description of ACN would suffer from being too structured without long-range corrections.

Finally, a generally lower QM/QM binding energy is obtained in LCAO mode, compared to the pure grid based Finite-Difference mode. This is attributed to the BSSE [38, 39] in the LCAO mode, where, at sufficiently short distances, the (incomplete) basis set of each ACN overlaps, so the effective basis of each molecule is larger (less incomplete) than the basis of each molecule at long distances. Thus, due to the variational principle, the potential energy of the dimer becomes lower at short distances, compared to what it would be, *without* the overlapping wave functions. Since these calculations only serve to rationalise the RDF-differences, no further BSSE-corrections (like the counterpoise correction [39]) has been employed.

Based on these observations, we have chosen to simulate the complexes of interest in a purely classical solvent. This will provide the best achievable solvation shell behaviour within the boundaries of the current version of the code. Furthermore, the exclusion of QM solvent molecules minimizes the scale of the QM subsystem, which is the computational bottleneck, and we avoid problems raised by QM solvent molecules diffusing out of the QM cell.

Of course, this choice also rules out studies of phenomena requiring full QM coupling of solute and solvent states, e.g. solvated electrons, (the acetonitrile equivalent of) photoaquation, etc. …—phenomena of high importance and interest, to simulators and experimentalists alike. However, the problem of switching the description of a solvent molecule diffusing out of the QM region is not trivial [40].

References

1. J. Mortensen, L. Hansen, K.W. Jacobsen, Phys. Rev. B **71**, 035109 (2005)
2. J. Enkovaara, C. Rostgaard, J.J. Mortensen, J. Chen, M. Dulak, L. Ferrighi, J. Gavnholt, C. Glinsvad, V. Haikola, H.A. Hansen et al., J. Phys.: Condens. Matter **22**, 253202 (2010)
3. C. Rostgaard, Online Notes (2010), https://wiki.fysik.dtu.dk/gpaw/paw_note.pdf
4. E.Ö. Jónsson, K.S. Thygesen, J. Ulstrup, K.W. Jacobsen, *Computational Approach to Electron Charge Transfer Reactions* (Lyngby, Denmark, 2014)
5. P.E. Blöchl, Phys. Rev. B **50**, 17953 (1994)
6. P.E. Blöchl, C.J. Först, J. Schimpl, Bull. Mater. Sci. **26**, 33 (2003)
7. CAMd, Electronic Structure Calculations With the GPAW code, https://wiki.fysik.dtu.dk/gpaw/documentation/literature.html. Accessed 1 Aug 2014
8. A.H. Larsen, M. Vanin, J.J. Mortensen, K.S. Thygesen, K.W. Jacobsen, Phys. Rev. B **80**, 195112 (2009)
9. A. Brandt, Math. Comput. **31**, 333 (1977)
10. J. Bernholc, D.J. Sullivan, E.L. Briggs, Phys. Rev. B **54**, 14362 (1996)
11. Y. Liu, D.A. Yarne, M.E. Tuckerman, Phys. Rev. B **68**, 125110 (2003)
12. A. Warshel, M. Levitt, J. Mol. Biol. **103**, 227 (1976)
13. H.M. Senn, W. Thiel, Top. Curr. Chem. **268**, 173 (2007)
14. H.M. Senn, W. Thiel, Angew. Chem. **48**, 1198 (2009)
15. A. Jaramillo-Botero, R. Nielsen, R. Abrol, J. Su, T. cal, J. Mueller, W.A.G. III. Top. Curr. Chem. **307**, 1 (2012)
16. M.W. van der Kamp, A.J. Mulholland, Biochemistry **52**, 2708 (2013)

17. The Nobel Prize in Chemistry 2013, http://www.nobelprize.org/nobel_prizes/chemistry/. Accessed 1 Sept 2014
18. A. Laio, J. VandeVondele, U. Rothlisberger, J. Chem. Phys. **116**, 6941 (2002)
19. M. Allen, D.J. Tildesley, *Computer Simulation of Liquids* (Oxford University Press, 1989). ISBN 0-19-855645-4
20. H.C. Andersen, J. Comput. Phys. **52** (1983)
21. L. Verlet, Phys. Rev. **159**, 98 (1983)
22. J. Gavnholdt, T. Olsen, M. engelund, J. Schiøtz, Phys. Rev. B **78**, 075441 (2008)
23. T. Olsen, J. Gavnholt, J. Schiøtz, Phys. Rev. B **79**, 035403 (2009)
24. E. Guardia, R. Pinzón, J. Casulleras, M. Orozco, F.J. Luque, Mol. Simul. **26**, 287 (2001)
25. N.L. Allinger, X. Zhou, J. Bergsma, J. Mol. Struct. **312**, 69 (1994)
26. H.A. Lorentz, Annalen der Fysik **248**, 127 (1981)
27. D. Berthelot, Comptes rendus hebdomadaires des séances de l′ Académie des Sciences **126**, 1703 (1898)
28. M. Waldmann, A.T. Hagler, J. Comput. Chem. **14**, 1077 (1993)
29. A.K. Al-Matar, D.A. Rockstraw, J. Comput. Chem. **25**, 660 (2004)
30. P.L.A. Polelier, A.J. Stone, D.J. Wales, Faraday Discuss. **97**, 243 (1994)
31. J. Siebers, U. Buck, T. Beu, Chem. Phys. **239**, 549 (1998)
32. E.M. Cabaleiro-Lago, J. Hermida-Ramón, A. Peña-Gallego, E. Martínez-Núñez, A. Fernández-Ramos, J. Mol. Struct. **498**, 21 (2000)
33. T.A. Ford, L. Glasser, Int. J. Quantum Chem. **84**, 226 (2000)
34. E. Schwegler, J.C. Grossman, F. Gygi, G. Galli, J. Chem. Phys. **121**, 5400 (2004)
35. T. Todorova, A.P. Seitsonen, J. Hutter, I.W. Kuo, C.J. Mundy, J. Phys. Chem. B **110**, 3685 (2006)
36. A. Møgelhøj, A.K. Kelkkanen, K.T. Wikfeldt, J. Schiøtz, J.J. Mortensen, L.G.M. Pettersson, B.I. Lundqvist, K.W. Jacobsen, A. Nilsson, J.K. Nørskov, J. Phys. Chem. B **115**, 14149 (2011)
37. I. Lin, A. P., I. Tavernelli, U. Rothlisberger. J. Chem. Theory Comput. **8**, 3902 (2012)
38. J.H. Jensen, *Molecular Modeling Basics* (CRC Press, 2010)
39. F. Jensen, *Introduction to Computational Chemistry*, 2nd edn. (Wiley, 2007)
40. R.E. Bulo, C. Michel, P. Fleurat-Lessard, P. Sautet, J. Chem. Theory Comput. **9**, 5567 (2013)

Chapter 6
Direct Dynamics Simulations of $Ir_2(dimen)_4^{2+}$

Figure 6.1 shows two of the main dynamic, structural modes for $[Ir_2(dimen)_4]^{2+}$. As described in Chap. 2, the dimen ligand provides the optimal compromise between flexibility and rigidity for large, but controllable structural changes e.g. by electronic excitation. Electronically exciting this complex promotes an electron from the antibonding HOMO to the bonding LUMO, effectively forming a chemical bond between the Ir atoms [1–3].

Here, we aim to expand on the pre-existing knowledge about the system by simulating an adiabatic version of the vibrational motion in the single molecule.

We approximate the experimental excitation by instantaneously promoting a comprehensive representation of the GS configuration space to the ES, neglecting effects from finite pulse-widths and bandwidths of real excitation sources (the effect of the bandwidth will be examined in Sect. 6.7. The ultrafast ES dynamics are believed to take place on S_1, but since similar binuclear d^8–d^8 complexes have previously been shown to have the same triplet and singlet surface shapes, only differing in energy [4, 5], it is assumed possible to carry out T_1 simulations mimicking the ES S_1 dynamics. Thus, the computational cost is kept within the feasible range for systems of this size by staying within the GS DFT framework on T_1. This approximation is implied for all the excitations simulated in this work.

By benchmarking the simulation results against the established experimental knowledge from Chap. 2, we can obtain information not experimentally distinguishable, such as possible dynamics of spectroscopically dark or obscured modes, direct observation of intra- and intermolecular energy transfer and solvent interactions.

First, however, it is necessary to ascertain the adequacy of the model description.

A.O. Dohn, *Transient Changes in Molecular Geometries and How to Model Them*, Springer Theses, DOI 10.1007/978-3-319-18747-1_6

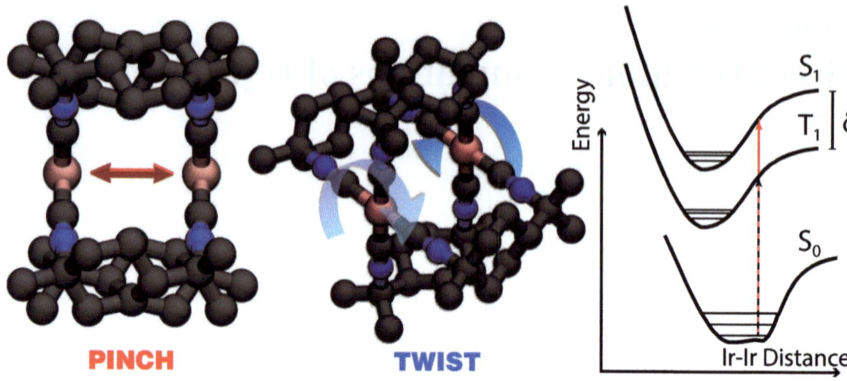

Fig. 6.1 *Left* Two of the main dynamic, structural modes of $[Ir_2(dimen)_4]^{2+}$, here shown without hydrogens for clarity, for the long and eclipsed conformer, as described in Chap. 2, p. 10 and on. *Right* Sketch of a Franck-Condon diagram, showing how experimental excitation is approximated in the simulations

6.1 Preparatory Tests

6.1.1 Cell Size

Figure 6.2 shows that a simulation box with 4 Å of vacuum padding in each dimension was enough to fully eliminate effects from truncating the wave functions, so the cell size for each simulation (both GS and ES) was chosen thusly.

6.1.2 Basis Sets and Grid Spacing

Test geometry calculations were performed in vacuum, and the results are shown in Fig. 6.3. The max Ir–Ir distance difference in the grid spacing region 0.15 Å $\leq h \leq$ 0.21 Å is less than 3.5 % of the largest distance, so instead of fixing the grid spacing to a definite value, it was possible to speed up the simulation even further by keeping the number of grid points to values divisible by eight, i.e. (112, 120, 112) for the (x, y, z) dimensions.[1] Since the overall size of the molecule varies for each of the initial GS configurations used in the the ES simulations, so does the QM cell, resulting in grid spacings 0.1571 Å $\leq h \leq$ 0.1931 Å, well within the converged region shown in Fig. 6.3. The basis set used was of tzp quality for Ir and dzp for the rest of the molecule. The chosen PBE vacuum description overshoots the Ir–Ir distance, when compared to the experimental value of the solvated complex [6]:

[1] This is an internal aspect of the parallelization in GPAW in the LCAO-mode, see https://wiki.fysik.dtu.dk/gpaw/documentation/lcao/lcao.html.

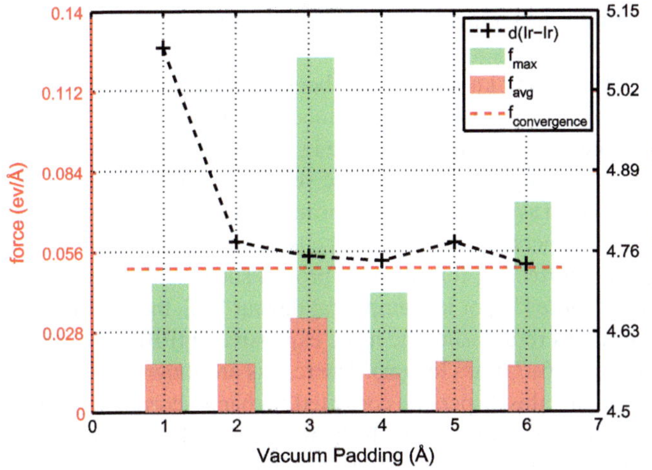

Fig. 6.2 Ir–Ir distance as a function of cell size. The relaxation does not converge for cells with 3 Å of vacuum, but for vacuum padding of 4 Å or more, the Ir–Ir changes are within the accuracy of the real space method. The calculation with the largest cell did not converge within the allowed walltime. The calculations were carried out using a grid spacing of 0.18 Å and a tzp/dzp basis set for Ir/the rest of the molecule

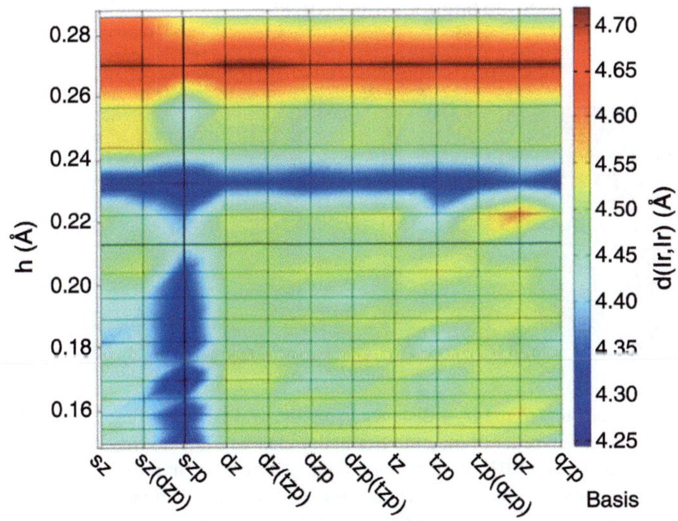

Fig. 6.3 Convergence of the main structural parameter of the complex—the Ir–Ir distance—with respect to the real space grid spacing h, and size of basis set. A stable value for the distance is maintained when employing values of $h < 0.22$ Å with basis sets of at least double-zeta size. The convergence criteria for each individual relaxation was a maximum force of 0.05 eV/Å on any of the atoms in the system

4.3 Å. Crystalline values range from 3.601 to 4.414 Å, depending on the type of counter ion [7].

It has, to the best of our knowledge, not been possible for any electronic structure method of the same level to more precisely model the Ir–Ir distance without resolving to mapping out the potential energy surface in the parameters of the pinch and twist, fixing one while scanning the other and vice versa (as seen in Fig. 2.6, Chap. 2, p. 13, and [8]): An option not available when performing direct dynamics simulations. Since it is believed that dispersion forces could be the cause of this problem [9, 10], and many attempts on correcting for this have proved successful on smaller molecules, using methods of more [11] or less empirical flavour [12], the preliminary structural convergence-test was rerun using the more expensive optPBE-vdW functional [13] (each SCF step takes 6–7 times longer). The results are shown in Fig. 6.4. The figure shows that employing this dispersion corrected exchange-correlation functional does not produce geometries with shorter Ir–Ir distances. While the result might seem surprising, it has been shown that similar dispersion-corrected functionals produce the correct geometries of other d^8–d^8 systems due to dispersion effects in the π-orbital stacking of the metal *ligands*, and that the actual *metal-metal* interactions only represent 10–15 % of the total dispersion contribution [10].

Since the use of this exchange-correlation functional did not increase the agreement with experimental value, was more expensive, and had convergence issues, the PBE functional was used throughout this project.

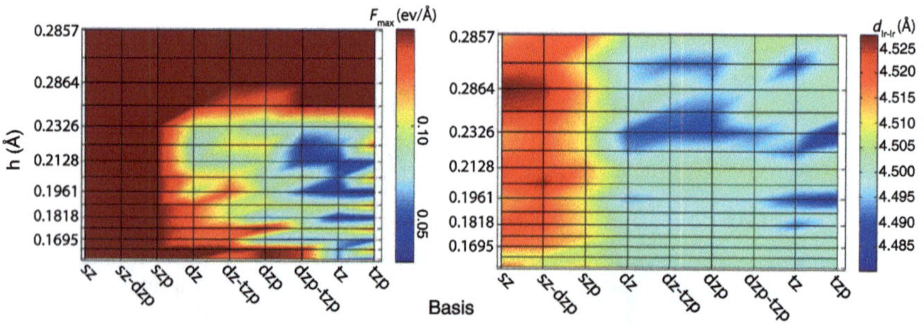

Fig. 6.4 Convergence test of the Ir–Ir distance using the optPBE-vdW functional. While the fluctuations in distance are an order of magnitude smaller than with the normal PBE, the convergence is not as convincing as with PBE. Furthermore, the mean distance in the most stable southeastern quadrant is still overshooting the experimental value. At last, this functional had many convergence problems, as can be seen from the plot of the average forces in the system, where only a few of the calculations reached forces smaller than the convergence criterium

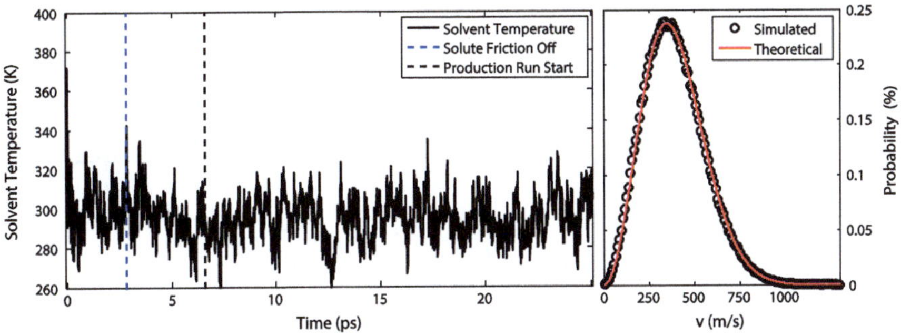

Fig. 6.5 *Left* Re-equilibrating the system after adding $[Ir_2(dimen)_4]^{2+}$. In first ~ 2.5 ps, the complex was also coupled to the Langevin thermostat. The part of the trajectory after the black stipled line was used as the production run for the ES simulations. *Right* The center-of-mass speed distribution of all ACN_S, sampled from the start of the production run. The avg. forces on the entire system is 2.36 μeV/Å per atom

6.2 Setting up QM $[Ir_2(dimen)_4]^{2+}$ in MM Acetonitrile

The QM region was defined as the $[Ir_2(dimen)_4]^{2+}$ complex, which was placed in the already equilibrated MM box of 28 × 28.5 × 31.5 Å, and 290 ACN molecules, giving the density of 0.786 g/cm^3. The ACN molecules overlapping the solute were then removed, resulting in a total of 237 solvent molecules. Figure 6.5 shows how the solvent equilibrates to the temperature set by the thermostat. The right side of the figure shows how the magnitude of the ACN center-of-mass velocities are distributed according to the Maxwell-Boltzmann distribution. The average temperature from the translational speeds is 300.6 K, and the average force per atom is 2.36 μeV/Å.

6.3 $[Ir_2(dimen)_4]^{2+}$ in the Ground State

Figure 6.6 shows the evolution of the main structural parameters in $[Ir_2(dimen)_4]^{2+}$ in the GS. The Ir–Ir distance oscillates reasonably within the k_bT region defined by van Driel [14] (see Fig. 2.6, p. 13). It could be argued that non-specific solvation can further stabilise the complex, lowering the overall energy, thus allowing for a somewhat larger oscillation amplitude. Furthermore, in calculating the energy curves, instead of changing the molecular structure along a normal mode reaction coordinate, all other intramolecular distances and angles than the parametrized Ir–Ir distance and dihedral angle were allowed to fully relax at each sampling point. Thus, what one could call an 'overly adiabatic' representation of the energetic landscape is obtained, in the sense that the *actual* motion of the atoms parametrized to obtain the surface might be too fast for the rest of the molecular structure to fully relax, when one of the normal modes of the molecule is activated. This method is now standard in

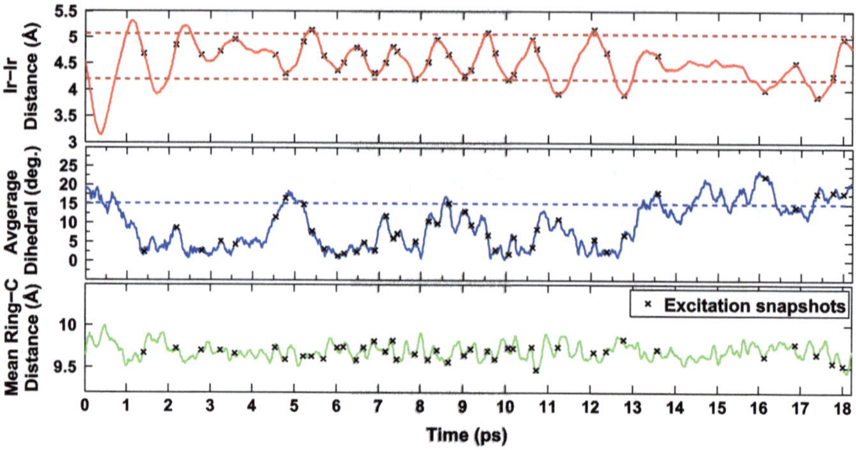

Fig. 6.6 Structural modes in $[Ir_2(dimen)_4]^{2+}$ plotted for the GS trajectory. *Top* Ir–Ir distance. *Middle* Mean dihedral angle. *Bottom* Mean distance between opposite carbon atoms in the ligand rings. This parameter describes a breathing motion which will play an important role in the excited state (see Sect. 6.4.3 and on). The stipled lines indicate the allowed region within $k_B T$ of the energy surfaces from [14] (see Fig. 2.6). The ES runs were started from the configurations marked with black crosses

computational packages like Gaussian, which has also been used to produce energy surfaces of $[Ir_2(dimen)_4]^{2+}$ [8].

When simply averaging the time between the peak positions during the time with the least amount of oscillation pertubations by solvent collisions (i.e. from 5 to 13 ps) a period of 1.2 ps is obtained, with a standard deviation of 0.2 ps. Averaging over the entire trajectory gives 1.5 ps. Hartsock et al. assign two GS oscillation modes with periods of roughly 3 and 0.7 ps from the long and short conformer, respectively. The average Ir–Ir distance from the simulation is 4.5 Å.

The difference in GS oscillation frequency, and also in the longer average Ir–Ir distance is ascribed to vdW-related DFT deficiencies in non-covalently-bonded interactions, previously discussed in Sect. 6.1.

There is no apparent correlation between the dihedral angle and the Ir–Ir distance.

The crosses in Fig. 6.6 show where the 40 GS configurations are excited, to get a total of 140 ps ES trajectories sampled for the ES production run. The sampling was spaced such as to minimize the correlation between each of the excited state trajectories, while also allowing for comparison of excitations from complexes of similar starting geometries, but with different solvent configurations, and excitations from different starting geometries. The thermostat was turned completely off for these runs. The ES timestep was 1 fs, to allow better resolution and minimize the energetic drift, which is analysed in Fig. 6.7.

To see the effect on the energy stability of turning off the thermostat completely, 4 ES runs were made with the thermostat still applied to the solvent, and the average

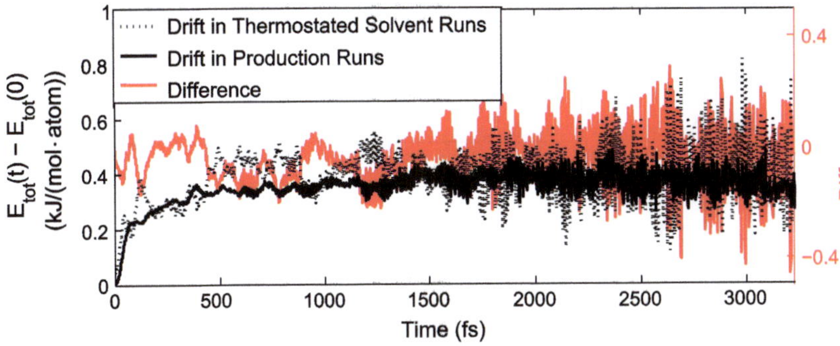

Fig. 6.7 Analysis of energetic drift, defined as the total energy $E_{tot}(t)$ minus the total energy at $t = 0$ after excitation, E_{tot}. The main energy increase is apparent in both un- and -thermostated simulations, which means that the observed increase in energy is due to the excitation

drift of those simulations are compared to the average drift in the production run of the 40 parallel simulations.

Turning off the thermostat entirely results in an average drift at the end of the production run of 0.40 kJ/mol pr atom. The maximum average drift is 0.50 kJ/mol per atom in the system, over the entire simulation duration. For comparison, the bond dissociation energy of the first hydrogen from a methyl group is 439 kJ/mol.

6.4 The Excited State Population of $[Ir_2(dimen)_4]^{2+}$

A visual inspection of the orbitals and electronic density from one of the 40 ES simulations in Fig. 6.8 confirms that approximating the experimental excitation on the triplet surface results in the expected Ir–Ir contraction, at face value. In the

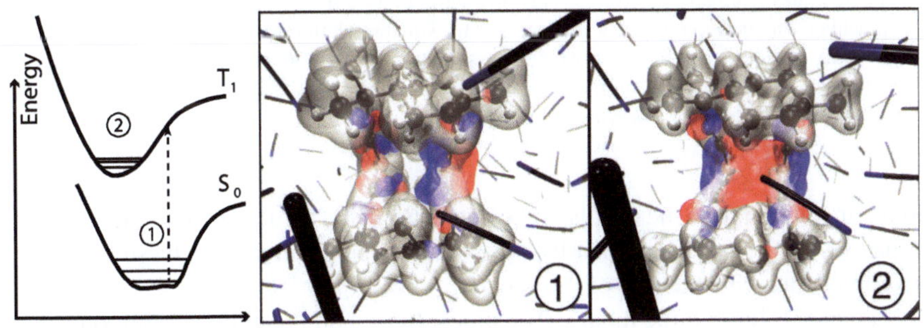

Fig. 6.8 Visualization of the simulation. The *blue* and *red* isosurfaces depict the HOMO in each case, and the grey isosurface is the total electronic density of the QM subsystem. All isovalues are identical between the two snapshots. *1* The complex in the ground state. The GS HOMO does not have any spacial overlap between the metal atoms. *2* A snapshot of one of the ES trajectories, at the point in time where the Ir–Ir contraction is largest. Here, we observe overlap of both the HOMO (the GS LUMO) and the resulting total electronic density

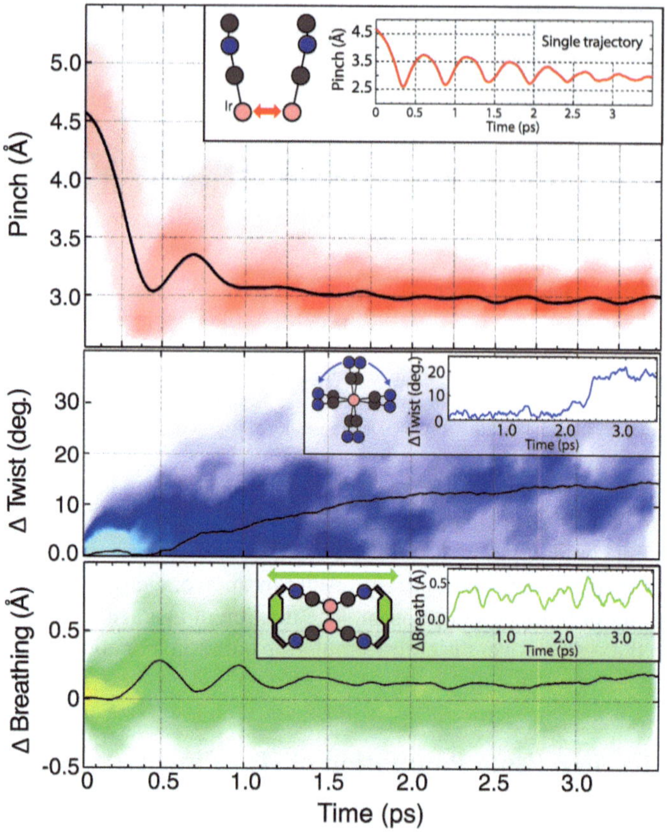

Fig. 6.9 The Ir–Ir pinching (*top*), twisting (*middle*), and breathing (*bottom*) as a function of time after excitation. The twist and the breathing is defined as the change in mean dihedral angle and in mean distance between opposite ring-carbons in the ligands, respectively, from the moment of excitation. The *black curves* are the averages, while the shaded backgrounds show the binned counts of values from all trajectories. The bin sizes are 50 fs by 0.1 Å, 0.05 Å, and 2 degrees for the pinching, breathing and twisting, respectively. This population loses its pinch phase within a picosecond, while the single trajectory-pinch oscillation dies out much slower, as can be seen on the top inset. However, the mean phase of the breathing mode is sustained for almost 1.5 ps. The pinching- and breathing modes share the same oscillation period

following, we analyse the results of the simulation, first by delving into the structural dynamics of the complex, then by investigating the coherence of these dynamics, and finally describe the role of the solvent.

Figure 6.9 shows three modes: The pinch, twist, and breathing (definitions given in the caption). The mean Ir–Ir distance in the population when measured over the last ps of the simulation is 2.98 Å, which is in very good agreement with previous experimental results from X-ray scattering [6] of 2.90 Å. When the Ir–Ir bond is

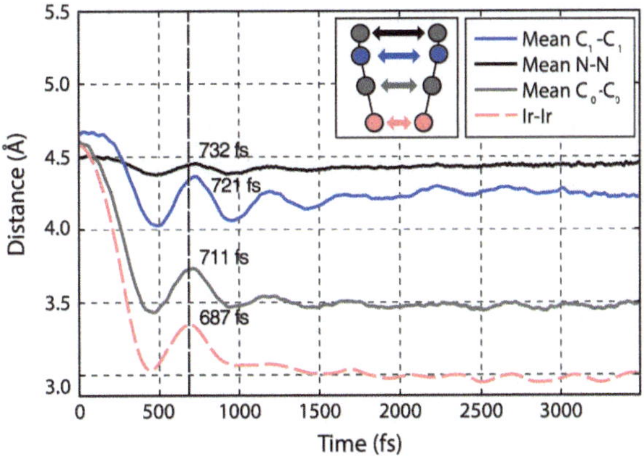

Fig. 6.10 The change in distance of the atoms oppposite to each other, adjacent to Ir. The two atoms in each ligand closest to Ir are also pinching, while the next carbon (C_1) is almost unaffected

formed in the ES, the accuracy of the DFT description no longer suffers as much from the poorly modelled vdW interactions.

The mean values of the entire population show oscillatory features, meaning that the motion of each trajectory is in phase with the others, thus showing signs of coherent motion, represented here in the average of the single trajectories.

6.4.1 The Pinching Mode

The excited population loses its mean pinch oscillation amplitude almost before completing a full period, but if the first oscillation period is calculated by doubling the time from the first well to the first top, the period is ~480 fs, or ~70 cm^{-1}. The frequency differs only 7 % from the experimentally observed frequency of 75 cm^{-1} [3]. This agreement further validates the assumption of very similar S_1 and T_1 PES shapes.

Figure 6.10 shows that the pinching motion is limited to the center part of the molecule, affecting the next two atoms adjacent to Ir, but not the outermost parts of the ligands. The ES bond formed between the Ir is the main instigator for the activation of this mode, since the oscillation peak in the neighbouring atom-pairs is systematically delayed with increasing distance from the metals. The onset of the N–N pinch is delayed about 200 fs, but the phase is maintained for roughly 1.5 ps, longer than the Ir–Ir pinch. This behaviour would not be modelled correctly on the previously discussed type of surfaces where all other atoms than the parametrised are allowed to fully relax at each sampling step.

The anharmonicity of the underlying potential is evident in the direct dynamics from the oscillation in the single trajectory (top inset in Fig. 6.9): It reflects a steeper potential for the short distances, and gentler slope for the long ones. The oscillation becomes more sinusoidal with time, as the trajectory moves closer to the harmonic approximation limit, deeper in the potential. Similarly, the frequency increases as the molecule dissipates its excess vibrational energy (also observed for the single trajectory), which is equivalent to anharmonic potentials with Morse-like characteristics. Since the experimental frequency is first measured at times $t \geq 0.5$ ps after excitation (when the molecule has already dissipated some of the excess energy), it is expected that our first simulated oscillation period is longer than the experimental result.

6.4.2 The Twisting Mode

Also consistent with experimental results [3], little to no coherent twist oscillation is observed from excitations that start from the long, eclipsed conformation. However, a delayed twist mechanism is observed for the average motion, with the molecule starting to twist after roughly 500 fs. The delayed onset of the twisting suggests that the mode is activated by the pinching of the Ir atoms.

6.4.3 The Breathing Mode

The ES breathing amplitude in the individual trajectories can be roughly 4 times larger than the amplitude of the GS oscillation shown in Fig. 6.6, meaning that this mode is also activated by electronic excitation. The mode, like the twist, also has a delayed onset. It can also be suggested that it plays a role in extending the experimental coherence lifetime. The breathing mode is coherent for almost 1.5 ps. Furthermore, the frequency of the pinching and breathing is similar: The period from the first to the second top is ~450 fs, or ~74 cm^{-1}, only 1.3 % slower than the experimental pinch.

Since the breathing mode is localized in parts of the ligands closest to the solvent, it can be expected to be coupled to the solvent dynamics, as will be further analysed in Sect. 6.6.

6.5 Coherence Decay

The causes of coherence decay can be either statistical or dynamical in nature, i.e. either a result of the difference in initial configurations from which the system is brought up to the excited state or of (stochastic) energy dissipation in each molecule, either through the solvent (EVR) or through internal degrees of freedom (IVR). The single trajectory in the top inset of Fig. 6.9 reveals that vibrational relaxation is much slower than the population mean coherence decay, and not very stochastic in nature. This observation can be solidified by selecting energetically narrow, phase-coherent distributions and then analyse the coherence time of these sub-populations.

A meaningful way to select these sub-populations is evidently through comparisons to the experimental excitation energy and bandwidth.

6.5.1 Approximating the Experimental Excitation Bandwidth

In principle, an instantaneous excitation corresponds to an infinite bandwidth. However, a more realistic representation of an experiment can be obtained by including an effective, finite bandwidth [15]. This can be included by estimating the effective bandwidth of our range of ES simulations in Fig. 6.11, and compare the pinch oscillation from different, relevant subsets with similar sampling statistics of the entire ES population in Fig. 6.12, to elucidate how this affects the coherence lifetime.

The simplest way to approximate the excitation energy is simply to subtract a single point energy calculation of the system in T_1 from a single point energy calculation in the S_0. This is of course a rough approximation. Since the T_1 surface is believed to have a similar shape to the S_1, but lower in energy [4, 5], utilizing ground state DFT on T_1 can be expected to introduce the error in the energy giving the largest deviation, and give too low excitation energies. The bandwidth, obtained

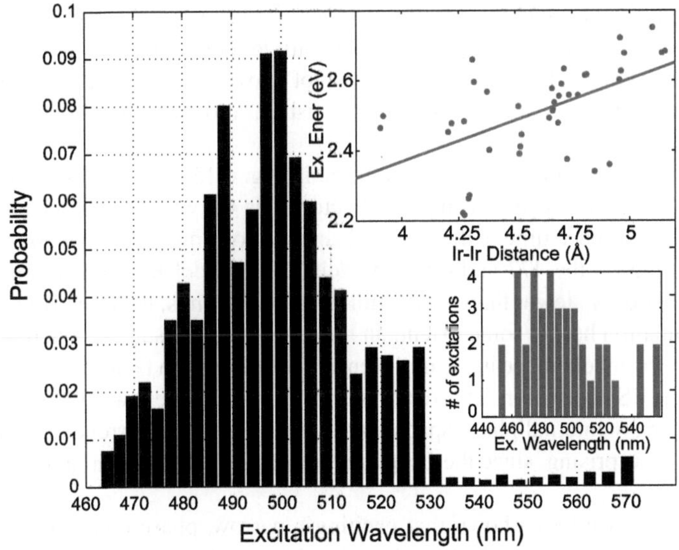

Fig. 6.11 Approximation of the effective bandwidth used in creating the excited state population. The excitation energy is approximated by subtracting the single point energy of each initial GS configuration from a single point calculation of its ES spin configuration counterpart. The *lower inset* shows histogram of these energies, converted to wavelengths, and how many times they appear in the ES population. This is extrapolated to the entire GS trajectory, under the (crude) assumption that the excitation energy is linearly dependent on (and only on) the Ir–Ir distance, shown in the *top inset*. This relation is then used for obtaining the results shown in the main figure

from looking at the excitation energies of all the excitations, should not be as sensitive to this approximation, since the error introduced should be the same for each of the excitations.

Due to the limits set by the computational costs, an extrapolation to the excitation energies of each step in the entire GS trajectory is made, although no clear linear correlation between the Ir–Ir distance and excitation energy is observed. The complex is known to have an absorption maximum assigned to the long and eclipsed conformer at 475 nm [3] (see absorption spectrum in Fig. 2.4, so it is expected due to the lower T_1 energy that the simulated excitation peak (Fig. 6.11, left) is located at lower energies. The experimental bandwith is 18.77 nm, and excitation wavelength is 477 nm [3]. The total effective bandwidth can be seen from the figure to be at least 100 nm, or 5 times the experimental bandwidth.

6.5.2 The Effect of the Bandwidth on Coherence Time

Figure 6.12 shows the Ir–Ir pinch in four subsets of the entire phase space of the ES population. The top two are subsets of excitations with energies corresponding to wavelengths of 477 and 494 nm, respectively. The first having the same value as in the experiment, and the second is lower in energy to account for the error made in simulating the excitation energies. The bottom left subset is chosen to have the most narrow internal phase space possible, with identical initial Ir–Ir distances, but different solvent configurations. The last plot shows the pinch in a wide internal phase space, but with similar sampling statistics, allowing for comparison of the four subsets.

As demonstrated in the lower right graph in Fig. 6.12, it is possible to obtain the same coherence-decay time using 12 trajectories as 40, which again means that any changes in coherence time in the other subsets are not simply due to worse statistics. Concentrating on the 477 and 494 nm excitation, the coherent pinch amplitude is observed to decay slower than for the full population. Thus, the effective bandwidth in the simulation has, to some extent, an influence on the resulting coherent motion. Therefore, the incoherence in the initial ensemble is the main factor for the coherence decay of the ES population. This result is underpinned by the lower left graph, where the very narrow internal phase space greatly increases the coherence lifetime. This result is not surprising, since the GS pinch oscillation has a similar amplitude as the ES pinch.

To sum up, we have shown that exciting a narrow, phase coherent distribution of GS configurations extends the coherence time on the excited state significantly. The analysis also shows that introduction of an effective bandwidth in making the ES population, similar to the one of the experimental excitation source, leads to a somewhat longer excited-state coherence time, although there is no strong correlation between the GS metal-metal vibrational amplitude/phase and the excitation energy.

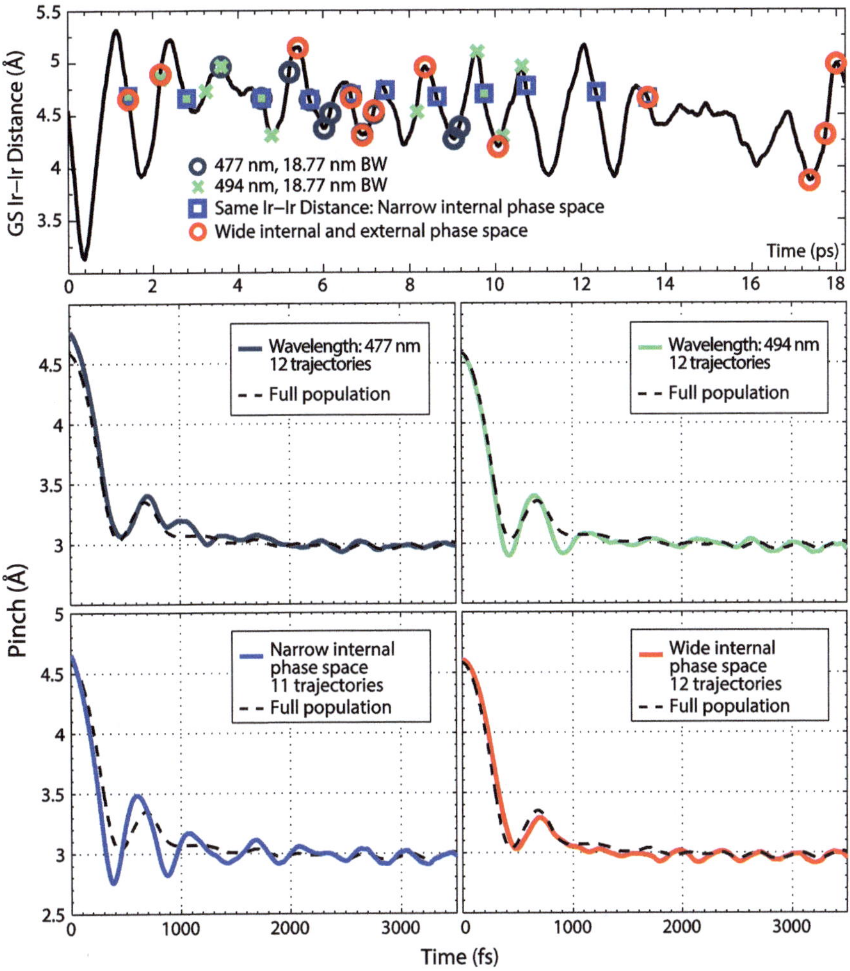

Fig. 6.12 Splitting up the total ES population with respect to excitation energies, or by grouping initial configurations. The *top* two ES pinch graphs are made from selecting for ES trajectories with excitation energies assumed close to the experimental value and bandwidth. The *bottom left* subset is made from a narrow internal phase space of initial Ir–Ir distances. The *last graph* shows the dynamics of a wide phase space, but with the same statistics as the other three subsets

6.6 Coherence Decay and Solvent Interactions

In order to further understand the effect of the solvent on the dynamics of $[Ir_2(dimen)_4]^{2+}$, Fig. 6.13 compares the pinch of the single trajectory from the inset of Fig. 6.9 to a group of additional simulations that have progressively more gas-phase character (from bottom to top in the figure).

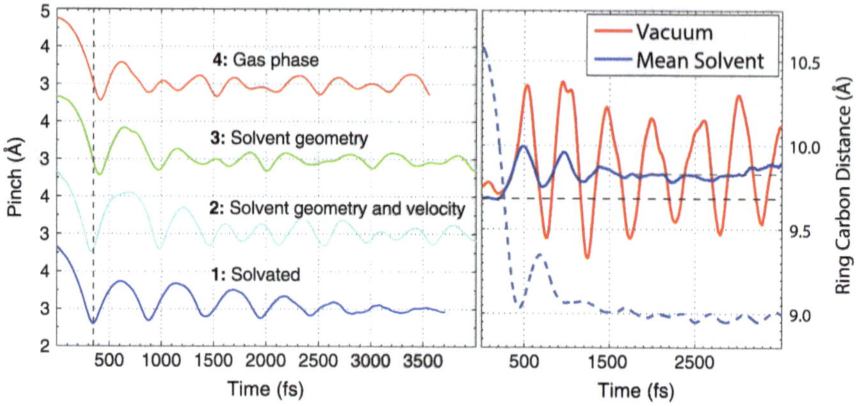

Fig. 6.13 *Left* Dynamics of the complex of progressively more gas-phase character (from *bottom* to *top*). *1* Pinch of the fully solvated complex. *2* A simulation where the solvent was removed at the time of excitation, but keeping the initial geometry influenced by the solvent, *and* the velocity of the atoms in the complex. *3* A simulation where only the geometry is kept, while the initial velocities are set to zero. *4* An excitation started from the gas phase, zero kelvin structure of the complex. *Right* Comparison of the average distance of the opposing ligands carbons that define the breathing mode. The average solvated pinch (*dashed blue curve*) is overlaid for timing comparisons. The *dashed black line* is the mean GS ring carbon distance

The first noticeable feature is the fast quenching of Ir–Ir oscillations in **4**, indicating efficient IVR in the pure gas phase simulation, which is evidently the *only* possible channel for energy redistribution in that system. In this case, the excess vibrational energy is efficiently transferred away from the pinch.

The original trajectory (**1**) was excited from a GS phase space area of already contracting metal atoms, which explains why the first contraction of the metal atoms in simulations **1** and **2** is faster than in **3** and **4**. The second contraction is fastest in **1**, and the amplitude of the pinching motion is the smallest, consistent with anharmonic oscillation, damped by the solvent. However, at longer times, the solvent cage apparently *facilitates* the oscillatory nature of the pinching motion in **1**, compared to the other trajectories. This is also supported by the comparison of breathing amplitude in the vacuum trajectory with the average amplitude from the production run (Fig. 6.13, right), which is much larger in the case where no solvent cage is present. The distance between the dashed black line in the figure and the mean solvent ring carbon distance indicate an average expansion of the distances between the opposing ligands of roughly 0.15 Å.

Figure 6.14 shows that not all solvent configurations block the breathing mode. The figure furthermore analyzes the interplay between IVR and solvation: In trajectory #1 on Fig. 6.14, the pinch energy is redistributed almost completely to the breathing mode within 1.5 ps, where solvent-induced interactions perturb the oscillation. Some of the remaining energy is then again transferred back into the pinch before dissipating further. In trajectory #2, the solvent configuration apparently does not allow for IVR to the breathing mode, so the excess excitation energy is contained

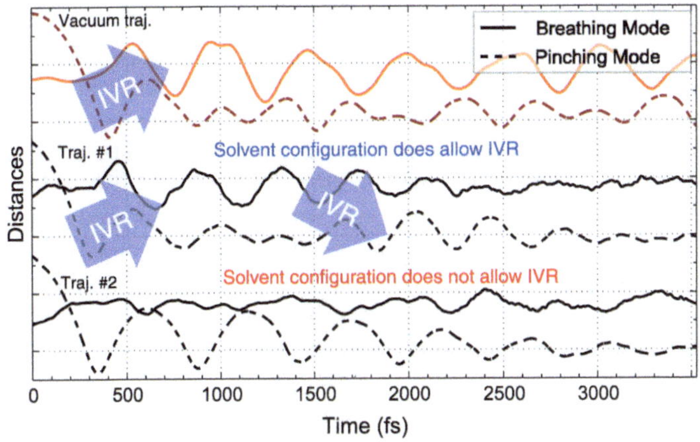

Fig. 6.14 Comparison of the breathing mode in the gas phase trajectory (*red*) and two trajectories of solvated systems, displaying the energy transfer mechanisms in the system. The two solvent-included trajectories are started from similar GS conditions, but 10 ps apart, so the two accompanying solvent configurations are uncorrelated. The breathing mode amplitude is large in gas phase, compared to the other trajectories, since there is no dampening from the solvent. The small pertubations in the breathing oscillation is attributed to solute-solvent interactions, governing the vibrational relaxation through the solvent

in the pinch, which has almost none of the smaller perturbations originating from interactions with the solvent.

All in all, this means that vibrational relaxation of $[Ir_2(dimen)_4]^{2+}$ at early times is substantially mediated by IVR into the breathing mode and that, in some instances, the solvent may hinder this motion which, in turn, extends the coherence time of the pinch compared to vacuum.

6.6.1 Comparison to Other Systems

That the coherence decay is faster than vibrational cooling is in contrast with the findings of van der Veen et al. [4] for the bimetallic d^8–d^8 complex $[Pt_2(P_2O_5H_2)_4]^{4-}$ in various solvents, where the authors observe that coherence decay occurs on the same time scale as vibrational cooling. While the two complexes are different, and different solvents are used (including protic solvents), they both have coordination sites along the metal atom axis, which could facilitate significant electrostatic- and dispersion interactions with the solvent. Nevertheless, the main difference causing the contrasting results is most likely due to the different rigidities of the two distinct ligand types, where the more flexible dimen ligand allows for a wider range of configurations in the GS ensemble. We emphasize that here, the variation of the GS metal-metal distance (Fig. 6.6) is comparable to the amplitude of the excited state vibration (Fig. 6.9).

Both the $[Ir_2(dimen)_4]^{2+}$ and $[Pt_2(P_2O_5H_2)_4]^{4-}$ experiments represent the formation of a chemical bond in a 'scaffolded diatomic'. Recently, collaborators have investigated the solvent-induced bond formation in a true diatomic molecule: The ground-state recombination of I_2, following photo-induced dissociation [16]. In this system, we observed the same behaviour as for $[Ir_2(dimen)_4]^{2+}$: The coherence decay is much faster than vibrational cooling. However, since I_2 is bound in the ground state, the original GS distribution is very narrow, compared to amplitude of the motion following photo excitation, and it is therefore not a source of decoherence. The cause of I_2 decoherence is still statistical: It is due to the direct EVR-interaction with its solvent cage, which lacks in both the order and rigidity compared to the molecular scaffold of the bi-metallic complexes.

6.7 The Solvent Shell Response

Due to the amount of sampled QM/MM statistics, it is possible to generate pairwise RDF_s from the Ir metals to the solvent atoms, and thereby observe how the solvent shell on average responds to the electronic excitation.

Figure 6.15 bins the 40 ES trajectories in 50 fs periods, and samples the RDF in each time step, resulting in a total of 200 images used for each ES curve, whereas the GS RDF was sampled over the entire GS trajectory.

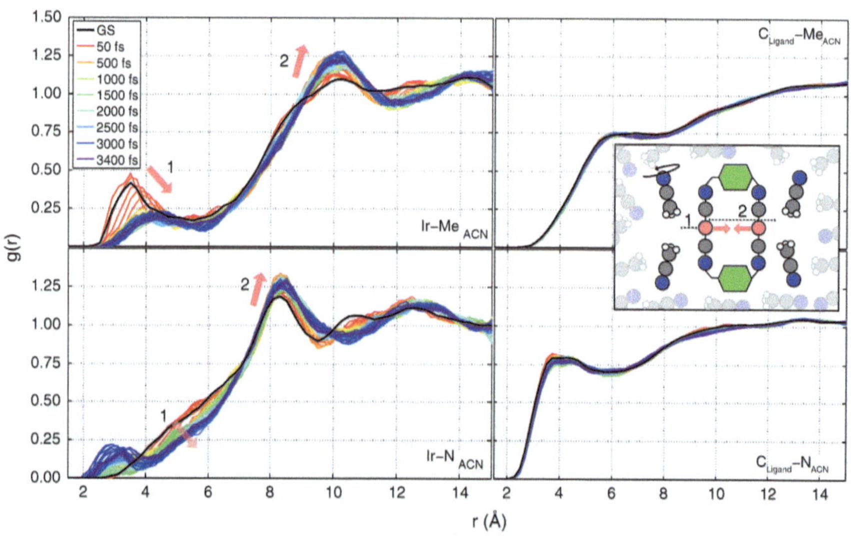

Fig. 6.15 Solvent-related RDF_s. *Left side* Between the iridium atoms and each end atom of the solvent molecules. *Right side* Between the dimen ligand carbons and each end of the solvent molecules. The RDF_s are sampled with $\delta r = 0.2$ Å at each time step over all the ES trajectories, and temporally averaged with a bin size of 50 fs, and finally smoothed with a 5 point moving average

In the ground state, the RDF of the ACN methyl groups and the Ir atoms show a peak around 3.5 Å (Fig. 6.15, top left). Due to the bulky dimen ligands (The distance from Ir to the end dimen carbons is between ~ 4.5 and ~ 7.5 Å), this peak must correspond to a pseudo-ordering of the methyl ends with respect to the solvent-accessible Ir-regions of $[Ir_2(dimen)_4]^{2+}$, parallel to the Ir–Ir axis, which in this analysis will subsequently be termed the 'side-regions' of $[Ir_2(dimen)_4]^{2+}$.

There is no analogue to the Ir–Me peak in the Ir–N RDF. In order to understand this Me–Ir pseudo-ordering, it is necessary to draw in observations from the RDF_s between the dimen carbons and the solvent (right side of Fig. 6.15). Still focusing on the GS, we also observe a steeper increase in probability of finding ACN nitrogens than methyl groups. Altogether, these observations can be explained if the ACN molecules in the $[Ir_2(dimen)_4]^{2+}$ side-regions prefer an orientation parallel to the complex, with the N ends oriented towards the ligands, and the Me ends towards the metals. This explains the steeper increase C_{ligand}–N probability. Finally, this explanation is supported by a Bader analysis [17] of a snapshot of the GS trajectory, which, naively speaking, integrates the computed electronic density back on the atomic centers. The analysis assigns roughly half a formal positive charge to each Ir atom, but over 1 negative formal charge to each of the ligand nitrogens, which are stabilized by the electro-positive methyl ACN ends.

It was necessary to employ a 5 point moving average smoothing filter on the resulting curves, which impedes any precise quantitative analysis of the results, but a clear change is seen in the ES distributions: There is a fast (<1 ps) decay and $\sim$0.7 Å displacement of the $g_{\mathrm{Ir-Me}}(r)$ peak, and an equally fast grow-in at $r > 9$ Å. It is highly unlikely that it is possible for structural changes to permeate 9 Å within a ps, thus these features are caused by the *metals* contracting, seen from the perspective of the solvent, and not the solvent itself responding.

The high-r regions are further investigated in Fig. 6.16, where the (unsmoothed) RDF_s has been averaged over the regions where changes take place, and the single RDF values have been plotted against time. The width of the ligands is approx. 5.5–6.5 Å. An oscillation in probability is seen for both the Ir–Me and Ir–N RDF. The decrease in probability in the 4–7 Å region of the Ir–N RDF corresponds to the metal contraction seen from the ACN Ns, which have the preferred orientation closer to the dimen ligands, away from the Ir atoms. This Ir–Ir contraction from the point of view of the solvent can also be visualised through the coordination numbers, as shown in Fig. 6.17.

Two regions are examined: **(1)** The short-r region, which is defined such that only coordinating solvents on the metal-ends of the complex are counted. This is done by choosing a maximum r distance shorter than the distance to the first solvent molecules coordinating to the dimen ligands. **(2)** The long-r region starts from the end of the short-r region and ends when the running coordination numbers $c_n(r)$ for all times are as numerically close to each other as possible, meaning that the bulk solvent shadows out the dynamics. Both $c_{n,\mathrm{Ir-Me}}(r)$ and $c_{n,\mathrm{Ir-N}}(r)$ (insets in Fig. 6.17) show an increase in slope (albeit the latter more subtly) that fits with the first RDF peak plus the Ir–Ir distance, meaning that here, the $c_n(r)$ starts also sampling solvent molecules on the other side of the complex. We can rule out sampling of

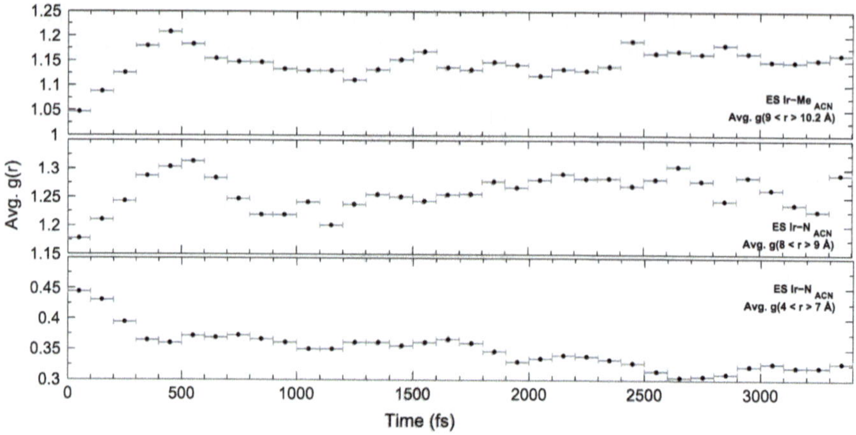

Fig. 6.16 The development of averaged RDF values over time. The time-binning has a width of 100 fs, and the r span is: *Top* Ir–Me r from 9 to 10.2 Å. *Middle* Ir–N from 8 to 9 Å. *Bottom* Ir–N from 4 to 8 Å

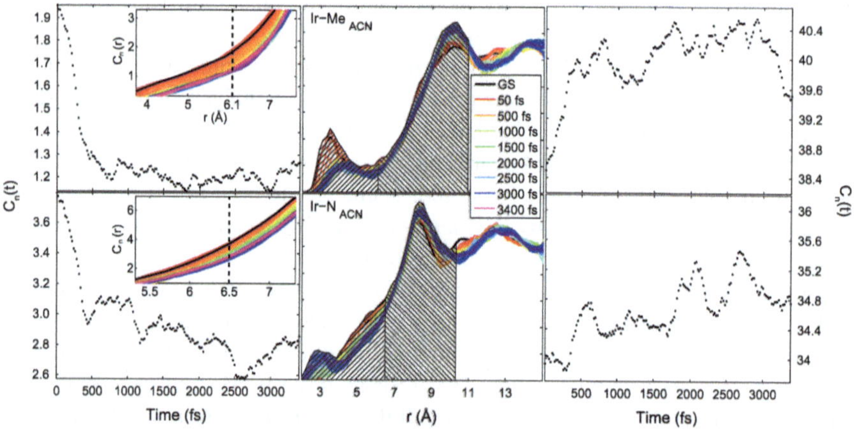

Fig. 6.17 The evolution of Ir-Solvent coordination numbers with time. The marked areas of the RDF_s in the middle graphs demarcate the regions used for the the plots of the temporal evolution of the short-r and long-r region (*left* and *right* plots), respectively. The *insets* show the running coordination numbers, $c_n(r)$ and the *dashed line* shows the end value chosen for the short-r region. The $c_n(r)$ slope increases around 6.1 Å for $c_{n,\text{Ir-Me}}(r)$, indicating that the $g(r)$ is now also sampling coordinating solvent molecules on the other side of the complex. For the less coordinated nitrogen ends, the slope change is more subtle, but estimated to 6.5 Å. The right region starts from the end of the left, and is chosen to end when all the $c_n(r)$s are as similar as possible, indicating the bulk region has been reached. The time-bins for averaging is 20 fs

solvent molecules coordinated to the ligands since the value of ligand-carbon RDF_s from Fig. 6.15 is 0 until 2 Å. The immediate decrease in $c_n(t)$ following excitation at the short-r region corresponds to the increase in the long-r region. Some of the

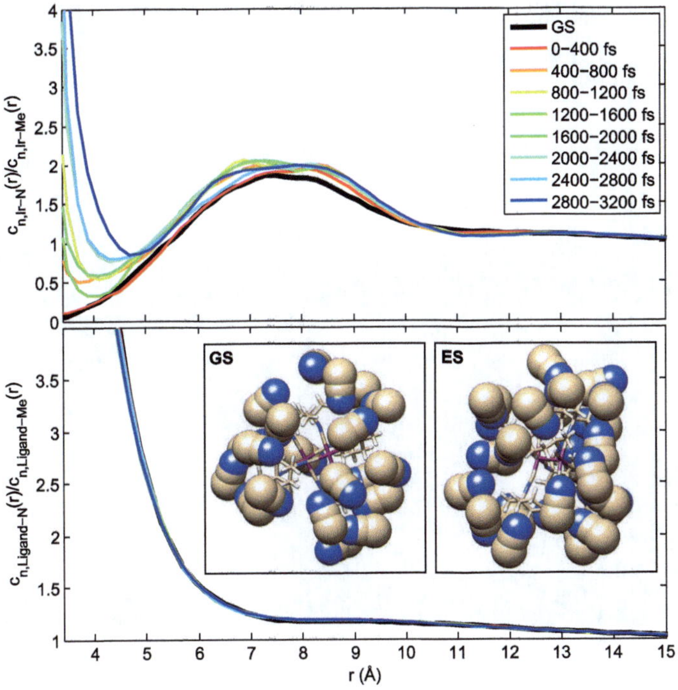

Fig. 6.18 Coordination number ratios of Ir–N:Ir–Me (*top*), and Ligand-N:Ligand-Me (*bottom*), averaged in 400 fs bins. For iridium, the GS ratio is below 1 for short distances, but this ratio significantly increases after the first 400 fs. For the ligands, no significant changes are seen, and the ratio always above 1. In the long-distance limit of the bulk solvent, the ratios converge to 1, which corresponds to the random orientation. The *insets* in the *bottom plot* depict snapshots of all the full ACN molecules within 4 Å of the solute

Ir–N coordination is recovered after 2.5 ps, even though the net coordination has decreased for this choice of volume.

The proposed model of the (changes in) pseudo-ordering of the solvent, debuted in Fig. 6.15, is further underpinned by Fig. 6.18. The figure shows the coordination number ratios of Ir–N:Ir–Me (top), and Ligand-N:Ligand-Me (bottom). A value of 1 is equivalent to a random orientation, since there is an equal amount of nitrogen atoms and methyl groups in ACN. In the GS, the Ir ratio at shorter distances is below 1, meaning that the solvent has a preferred orientation of pointing away from the Ir atoms. This changes significantly in the excited state trajectories, meaning that the preferred orientation changes as an effect of the excitation.

With time, the ACN-rotation shown in Fig. 6.18 settles, resulting in an Ir–N peak growing in around $r \sim 3$ Å (Fig. 6.15). The temporal evolution is more easily observed in Fig. 6.19, which is a surface plot of cut-outs of the r regions, evolving in time. In the top plot, the Ir–Ir contraction is again evident. At $t > 2$ ps, the Ir–N peak

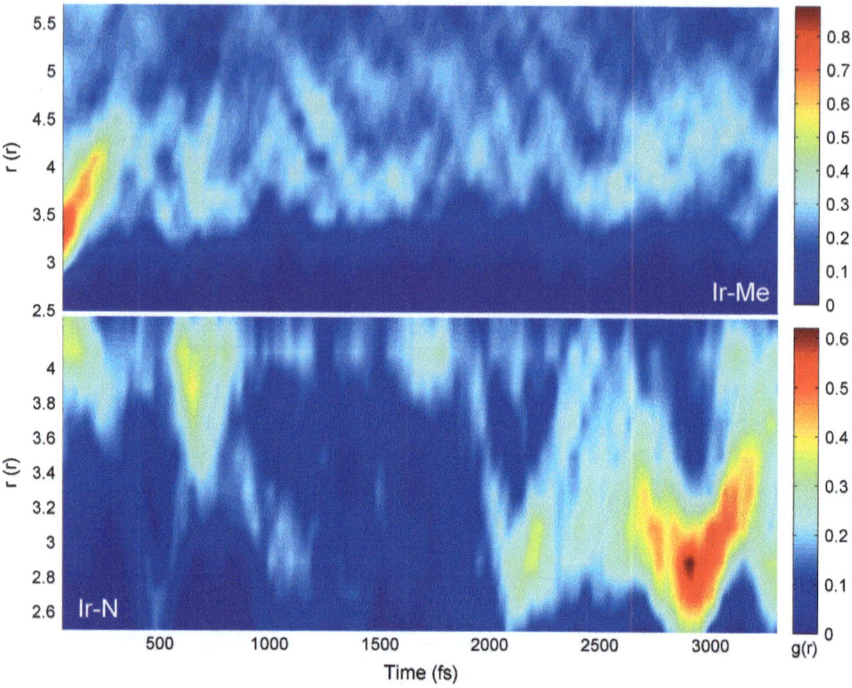

Fig. 6.19 2D plot of the temporal evolution of the RDF_s between Ir and each end of ACN. In the *top plot*, r for the peak increases with time, in conjunction with a decreasing $g(r)$-value. In the *bottom plot* the grow-in of a peak at short distances occur after roughly 2 ps

starts to grow in. This corresponds with the electronic charges on $[Ir_2(dimen)_4]^{2+}$ having moved into the center between the metal atoms, so the electronegative N-end can move closer to the Ir atoms, which effectively changes the solvent configuration according to the excitation.

It must be mentioned that the QM/MM ACN RDF_s in Fig. 5.9 of Sect. 5.5.1 showed potential overbinding features betweem the QM and the MM ACN_s. Even so, this does not necessarily have to carry over to the interaction of $[Ir_2(dimen)_4]^{2+}$complex with the MM solvent.

It is important to remember that the electronic excitation of $[Ir_2(dimen)_4]^{2+}$ does not change the direction of the dipole moment of the molecule, so it would have been tempting to assume that the solvent configuration remained completely unchanged, post excitation. As described in the above, the $[Ir_2(dimen)_4]^{2+}$ system presents a situation where such an assumption would be wrong.

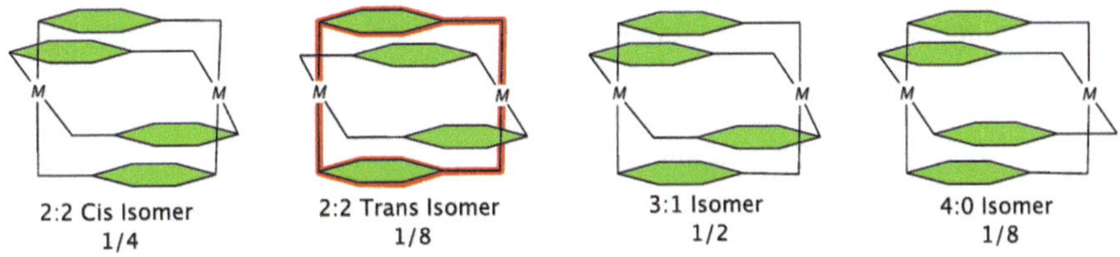

Fig. 6.20 The four $M_2(dimen)_4^{2+}$ Isomers as described in Sect. 2.2. The All simulations up until now has been carried out on the 2:2 Trans Isomer

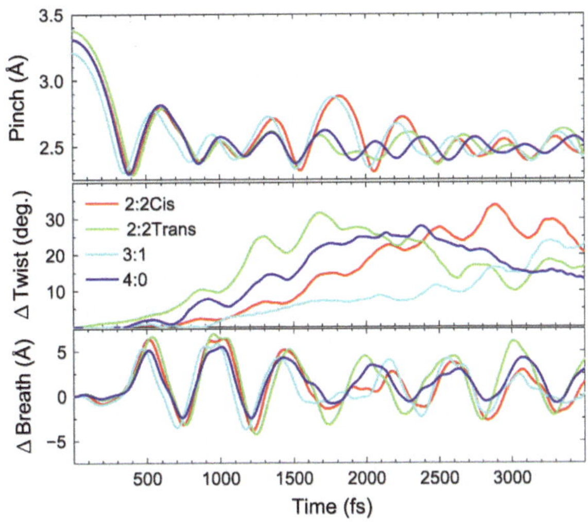

Fig. 6.21 Gas phase dynamics of all four $[Ir_2(dimen)_4]^{2+}$ isomers. All three main modes are expressed in all isomers, and the differences are at most on the same scale as the changes invoked by the different solvent configurations of the solvated simulations

6.8 Comparing Isomers

This section analyses the possible differences of having different ligand ring-orientations (Fig. 6.20) on the resulting dynamics. Gas-phase dynamical simulations with the same computational parameters as the pure gas-phase simulation on the 2:2 trans isomer of Sect. 6.6 was carried out, and the results are shown in Fig. 6.21.

While the initial Ir–Ir distance of gas phase-relaxed structures can vary up to 0.3 Å, all isomers show very similar gas-phase dynamical behaviour, where the differences are on the same scale as the differences invoked by solvating a population at 300 K. The 3:1 and 2:2 Cis isomers seem to recover some pinch amplitude after 1.5 ps, whereas for the others, the mode cools down quicker, and the average Ir–Ir distance at $t > 2$ ps varies from 2.99 to 3.03 Å.

The twist of the 3:1 isomer seems to have a longer onset and is generally slower. The other three isomers show weak signs of an overlying twisting oscillation with an average period of ~420 fs or 79 cm^{-1}, close to both the pinching- and the breathing period.

To sum up, re-orienting the ligands in the complex does not seem to change its internal dynamics more heavily than the solvent, and as such, comparing experimental data—which will be from a mix of the isomers—to the simulations purely on the 2:2 trans isomer should be completely feasible.

6.9 Conclusions

This section has demonstrated that the QM/MM implementation using GPAW for the QM description in our multiscale Direct Dynamics method can produce out-of-equilibrium molecular dynamics in statistical quantities that make it possible to obtain information on the average dynamics of populations, whereas many similar methods seem to be limited to produce results from single trajectories, due to computational costs.

With regards to the posed question of which modes channel the dissipation of energy, the experimental Ir–Ir pinch and twist dynamics have been reproduced, and additionally, we observed a, to our knowledge, not previously discovered breathing mode. This mode stays coherent for longer time than the pinch, and is apparently the first recipient of the excess energy transferred out of the pinch mode. We note that in some configurations, the rigidity of the solvent cage can block the breathing motion, and thereby sustain the coherence of the pinch mode.

By looking at the shapes of the individual trajectories, we have argued that the major electrostatic solute-solvent interaction occurs through the ligands for this complex, and that the IVR here is delayed, since the metals need to contract first. At last, the breathing occurs in the ligands, and is perpendicular to the pinch axis. As such, the (rigidity of the) ligands play an important role in the complex, both through defining the width of the GS ensemble, which affects the ES population coherence, and as mediators of the solvent interaction.

The investigation of the role of the solvent has shown how the Ir–Ir contraction is reflected in the correlation to the solvent, and thus note that any experimental X-ray scattering signal from the system will have an immediate contribution similar to an Ir–Me expansion, since the Ir atoms move inwards, away from the solvent. We have also shown that at later times, a non-specific solvent reorganization takes place, even though the dipole moment direction does not change.

Lastly we have argued from gas phase simulations of all isomers that the current simulational dataset will sufficiently describe the dynamics of the experimental mix of isomers. At the time of writing, the experimental data has just reached a stage where all the challenges of cleaning up the signal from an XFEL source have been met, and the future of the overall project looks promising. The next step will involve employing the signal-simulation techniques from Chap. 4 in conjunction with our experimental collaborators.

References

1. K.R. Mann, J.G.I. Gordon, H.B. Gray, J. Am. Chem. Soc. **97**, 3553 (1975)
2. M.D. Roundhill, H.B. Gray, C. Che, Acc. Chem. Res. **22**, 55 (1989)
3. R.W. Hartsock, W. Zhang, M.G. Hill, B. Sabat, K.J. Gaffney, J. Phys. Chem. A **115**, 2920 (2011)
4. R.M. van der Veen, A. Cannizzo, F. van Mourik, A.J. Vlcek, M. Chergui, J. Am. Chem. Soc. **113**, 305 (2011)
5. A.E. Stiegman, S.F. Rice, H.B. Gray, V.M. Miskowski, Inorg. Chem. **26**, 1112 (1987)
6. K. Haldrup, T. Harlang, M. Christensen, A. Dohn, T.B. van Driel, K.S. Kjær, N. Harrit, J. Vibenholt, L. Guerin, M. Wulff et al., Inorg. Chem. **50**, 9329 (2011)
7. C.L. Exstrom, D. Britton, K.R. Mann, Inorg. Chem. **35**, 549 (1996)
8. B.M. Hunter, R.M. Villahermosa, C.L. Exstrom, M.G. Hill, K.R. Mann, H.B. Gray, Inorg. Chem. **51**, 6898 (2012)
9. J. Klimeš, A. Michaelides, J. Chem. Phys. **137**, 120901 (2012)
10. S. Grimme, J. Djukic, Inorg. Chem. **50**, 2619 (2011)
11. S. Grimme, J. Comput. Chem. **25**, 1463 (2004)
12. S. Grimme, J. Antony, S. Ehrlich, H. Krieg, J. Chem. Phys. **132**, 154104 (2010)
13. J. Klimeš, D.R. Bowler, A. Michaelides, J. Phys.: Condens. Matter **22** (2010)
14. T.B. van Driel, M.M. Nielsen, *Time-Resolved X-ray Scattering of Molecules in Solution: Approaching the Molecular Movie* (Lyngby, Denmark, 2014)
15. J. Petersen, N. Henriksen, K. Møller, Chem. Phys. Lett. **539–540**, 234 (2012)
16. J.H. Lee, M. Wulff, S. Bratos, J. Petersen, L. Guerin, J.C. Leicknam, M. Cammarata, Q. Kong, J. Kim, K.B. Møller, et al., J. Am. Chem. Soc. **135**, 3255 (2013)
17. W. Tang, E. Sanville, G. Henkelman, J. Phys.: Condens. Matter **21**, 084204 (2009)

Chapter 7
Direct Dynamics Simulations of the Ru=Co Complex

This section describes the ongoing work on simulating structural- and solvation changes in the bi-centered Ru=Co complex described in Sect. 2.4.

7.1 Preparatory Tests

Figure 7.1 contains the various labels for the relevant geometric parameters under survey to the left, and the definitions of the electronic configurations to the right. The molecule spans almost 21 Å from left to right in the figure, and about 9 Å in height, in the depicted orientation. The Ru=Co 5+ ion contains 592 electrons (24 more than $[Ir_2(dimen)_4]^{2+}$), and its large spatial extent demands a larger QM cell than in the previous sections, increasing the computational costs. Three overall electronic states of the system have been studied:

- The electronic ground state: $[^1Ru^{II}{=}^1Co^{III}]$, GS.
- The proposed intermediate [III] $[^2Ru^{III}{=}^2Co^{II}]$, here labelled LS.
- The final excited state: $[^2Ru^{III}{=}^4Co^{II}]$, HS.

Many articles understandably define the GS as LS, since it *does* have the lowest total spin of *all* states: $[^1Ru^{II}{=}^1Co^{III}]$, but here we need to discern between all three states, so the low spin-label will be used on the *excited* state with the lowest spin.

7.1.1 Basis Sets and Grid Spacing

For the Ru=Co simulations, the QM cell size was chosen such that the molecule would be 'vacuum-padded'[1] with 5 Å in each direction, leading to a 25 × 20 × 30 Å cell.

[1]Meaning that the cell is 5 Å longer than the complex, in each direction, to avoid wave function truncation. The MM molecules are can of course still enter the QM cell.

A.O. Dohn, *Transient Changes in Molecular Geometries and How to Model Them*, Springer Theses, DOI 10.1007/978-3-319-18747-1_7

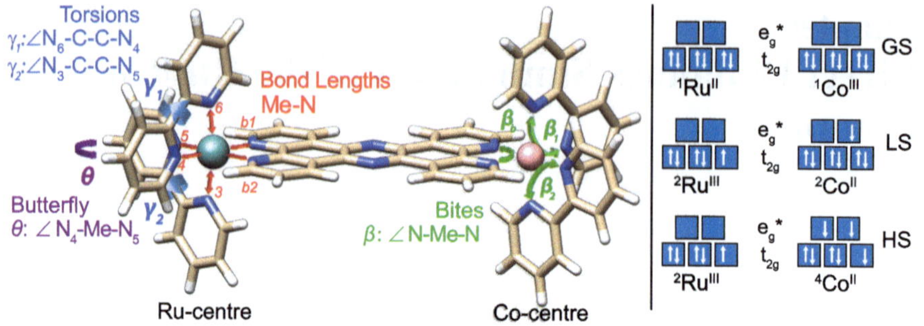

Fig. 7.1 *Left* Definitions of the relevant geometric parameters for the Ru=Co complex. The Ru-centre is to the *left*, and the Co-centre to the *right* side of the image. 'Me' (Metal) is either Ru or Co. While the bite angles are only depicted for the Co-centre, they are of course defined likewise for the Ru-centre, and vice versa for the torsions, bond lengths and butterfly angle. *Right* Labeling of the three electronic configurations simulated in this work

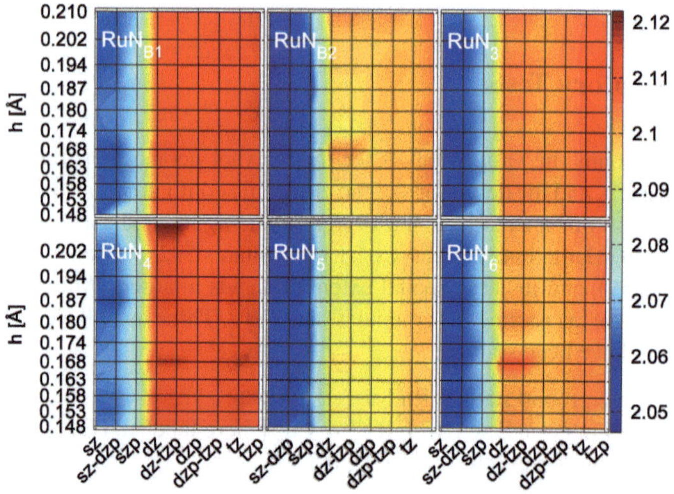

Fig. 7.2 Convergence of the Ru–N distances as a function of grid spacing h and basis set size. The *color bar* shows the distance in Å

Figures 7.2 and 7.3 show the convergence of the GS Ru/Co–N distances with respect to grid spacing and basis set size. In both cases, the major influence on the structure arises from the choice of basis set, which converges rapidly when basis sets of size larger than double zeta are chosen. For all values of $0.15\ \text{Å} < h < 0.20\ \text{Å}$, the variation in bond distance is at least 17 times smaller than the changes arising from changing the basis, meaning that all grid spacings under 0.20 Å will be acceptable.

Based on this analysis, we chose a grid spacing of $h = 0.18$ Å and dzp basis, with tzp basis on the metal atoms for all further simulations.

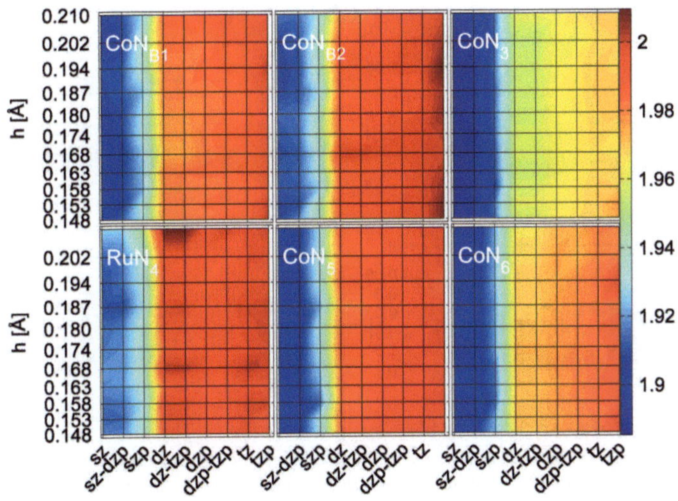

Fig. 7.3 Convergence of the Co–N distances as a function of h and basis set size. The *color bar* shows the distance in Å

7.1.2 Vacuum Structural Parameters for the Relevant Electronic States

We obtained the initial geometries from our collaborators M. Pápai and G. Vankó of the Wigner Research Centre for Physics, Hungarian Academy of Sciences. Their structure has been relaxed in vacuum, using the BP86 [1, 2] functional, which has the same exchange term as BLYP, but uses an older correlation term called Perdew86 [2], and is a first generation GGA-type functional, whereas PBE can be called second-generation [3].

Figures 7.4 and 7.5 show the quick convergence of the main structural parameters, namely the Me–N bonds, and are in principle somewhat trivial. Their purpose is mostly a contrast to the intramolecular instabilities of the unsolvated excited state structures, shown for the LS state in Figs. 7.6 and 7.7, and in Figs. 7.8 and 7.9 for the HS state.

The standard convergence criterion of F_{max} = 0.05 ev/Å is not reached within 300 steps. However, the average forces on the entire molecule converges around 0.1 ev/Å, and the bond lengths also seem to reach a stable value, even though the forces are still large. Especially the nitrogen atoms experience large forces, both in the LS and HS states. This indicates that the charge that has been moved by changing the magnetic moment on the metal atoms to fit the spin states of the molecules needs stabilization, e.g. from the solvent. The rest of the relevant structural parameters (torsion, butterfly angle and bite angle) show the same features, and can be found in Appendix A.

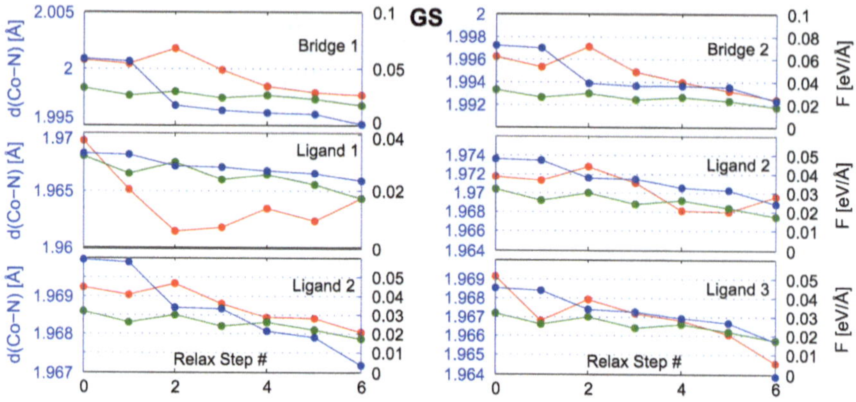

Fig. 7.4 Vacuum relaxed GS Co–N bond lengths (*blue lines*). The *green* graph represents the average force on the molecule, and the *red* is the average forces on the nitrogen atoms. For the GS, Convergence is reached within 6 steps, and the bond length changes are maximum 1 pm

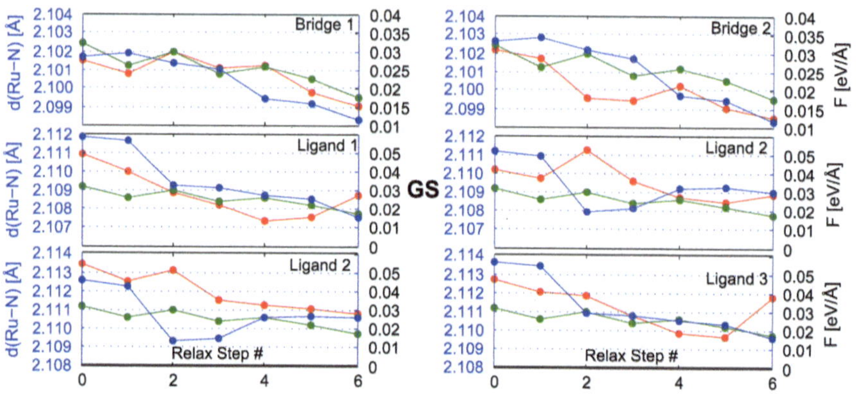

Fig. 7.5 Vacuum relaxed GS Ru–N bond lengths (*blue lines*), with the *red* and *green curve* displaying the avg. forces on the nitrogens and the entire molecule, respectively

Finally, for the spin-polarised calculations, GPAW suffered from significant SCF-convergence issues, regardless of whichever density-mixing scheme was attempted. Further attempts to correct this by increasing the number of empty states to include in the calculation or by adjusting the accuracy of the Poisson-solver were unsuccessful. Finally, convergence of the SCF-optimization could be recovered by 'smearing' of the occupation numbers using a Fermi-Dirac distribution with a width of 0.1 eV. The smearing does of course not take place in a real molecular system, but in order to run QM/MM MD simulations, it is crucial to employ a strategy that ensures convergence of the wave functions and the corresponding electronic density for each time step. And due to computational cost, it is also necessary to reduce the number of SCF-cycle steps as much as possible.

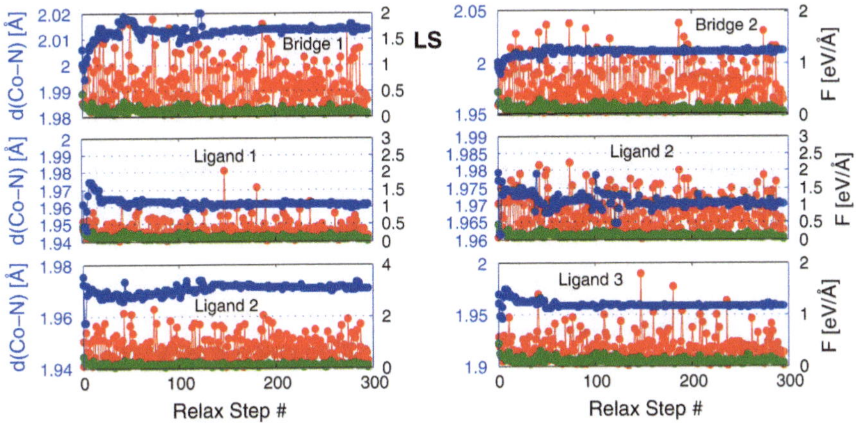

Fig. 7.6 Vacuum relaxed LS Co–N bond lengths. While there is a constant strain on the nitrogen atoms, the distances seem to converge to constant values

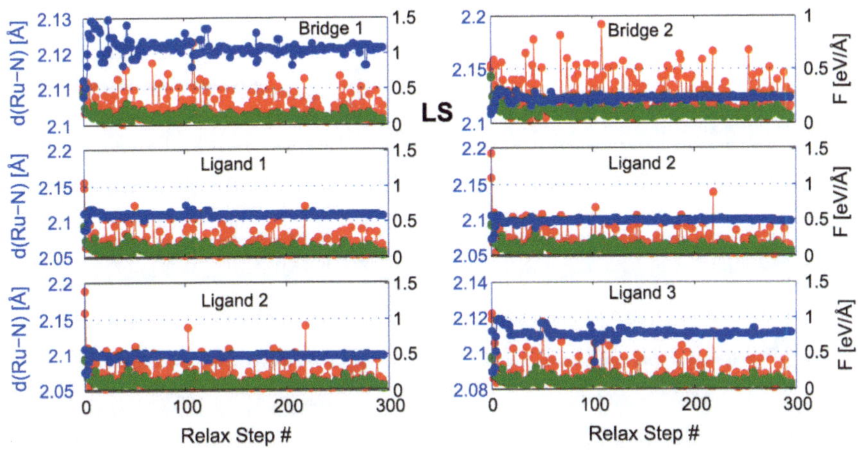

Fig. 7.7 Vacuum relaxed LS Ru–N bond lengths, where the same trend as for the Co–N bonds is observed

Based on this analysis, we have chosen the values at the last steps of the geometry optimizations as our 'vacuum values' for comparison with the (thermal distribution of) solvated bond lengths discussed in Sect. 7.3. The values can be found in Table 7.1. The obtained structural parameters reproduces part of the expected Co–N bond elongation for the HS, state, which in turn also changes the torsion angles γ, in the cobalt centre. The Ru-centre remains largely unchanged by the excitations, apart from a slight increase in torsion in the HS state.

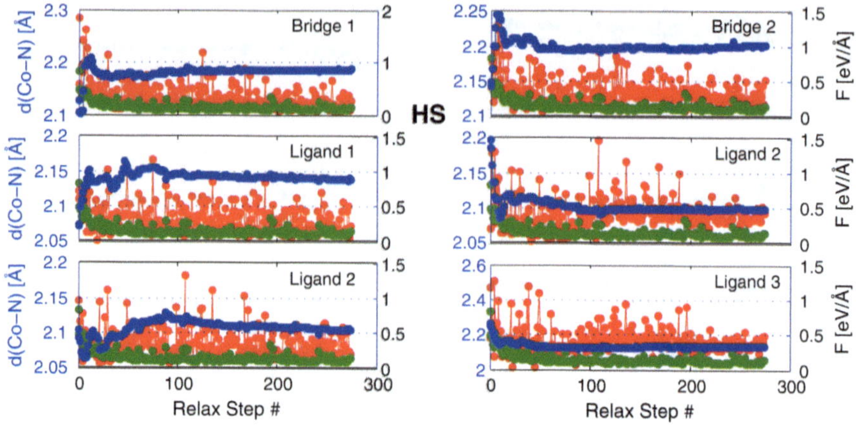

Fig. 7.8 Vacuum relaxed HS Co–N bond lengths, showing the same behaviour as the LS state. The Co–N_B bonds seem to have reached a stable distance, as well as the Co–N bonds 3 and 6. 2 and 3 seem to be close to convergence

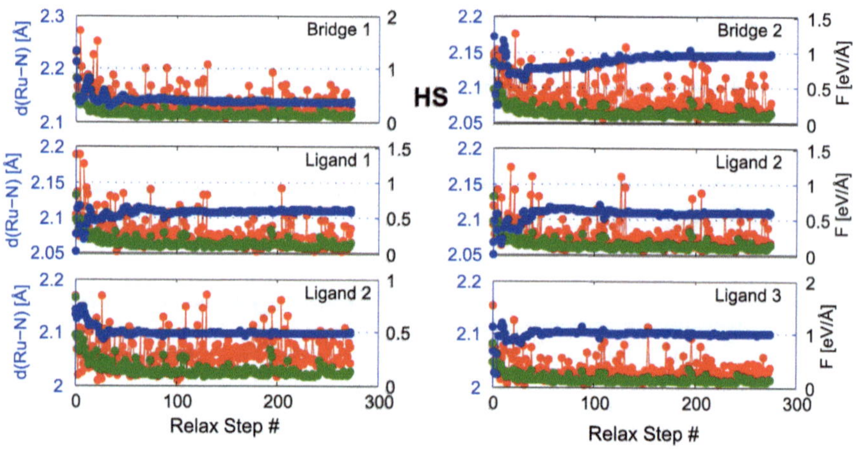

Fig. 7.9 Vacuum relaxed HS Ru–N bond lengths

7.2 Making the QM/MM System

For simulating Ru=Co the gas-phase geometry optimized GS structure was placed in a 32 × 32 × 44 Å box of pre-equilibrated acetonitrile, using the same method as with $[Ir_2(dimen)_4]^{2+}$. In this simulation, we opted to simulate two parallel GS trajectories:

- A trajectory with constrained H atoms, as in the previous section: Run C
- A trajectory with unconstrained hydrogens, but with twice the mass of a normal hydrogen: Run H

Table 7.1 Collected structural parameters for Ru=Co in gas phase

	Co-Centre			Ru-Centre		
	GS	LS	HS	GS	LS	HS
Me–N_{b1} (Å)	2.000	2.014	2.186	2.098	2.121	2.135
Me–N_{b2} (Å)	1.992	2.012	2.201	2.098	2.123	2.147
Me–N_3 (Å)	1.966	1.961	2.138	2.108	2.109	2.111
Me–N_4 (Å)	1.969	1.970	2.099	2.109	2.099	2.111
Me–N_5 (Å)	1.967	1.971	2.105	2.111	2.099	2.097
Me–N_6 (Å)	1.967	1.958	2.133	2.110	2.111	2.100
θ°	88.606	90.094	89.694	87.847	88.887	87.963
γ_1°	1.476	1.702	7.043	0.427	−0.943	−3.370
γ_2°	1.407	1.420	11.077	0.222	−0.998	−2.145
β_b°	81.972	82.158	75.253	77.792	78.331	77.048
β_1°	82.396	82.788	79.228	77.522	78.503	79.007
β_2°	82.295	82.727	77.688	77.535	78.488	78.322

The Co–N bonds elongate in the HS state, as expected. For the LS state, the Co–N bonds do not seem to be affected in a systematic manner. The Ru–N bonds are not significantly effected by the excitation. The HS torsion angle is also increased

Run H was made to (1) obtain a fully unconstrained thermally averaged structure which can be put into classical MD simulations for better statistics on solvation shell sampling as proposed in Sect. 4.3, and (2) to test the effect this has on the other structural parameters in the complex. Both simulations contain a total of 436 classically described ACN molecules. The thermostat is kept on the solvent part of the system during the entire simulation, to avoid drifts over these long, single trajectories.

Figure 7.10 confirms the stability and expected thermodynamic behaviour of run Cafter letting the solvent molecules re-equilibrate in the presence of Ru=Co, which happens within a ps, as is shown on the left inset. Figure 7.11 shows the equillibration

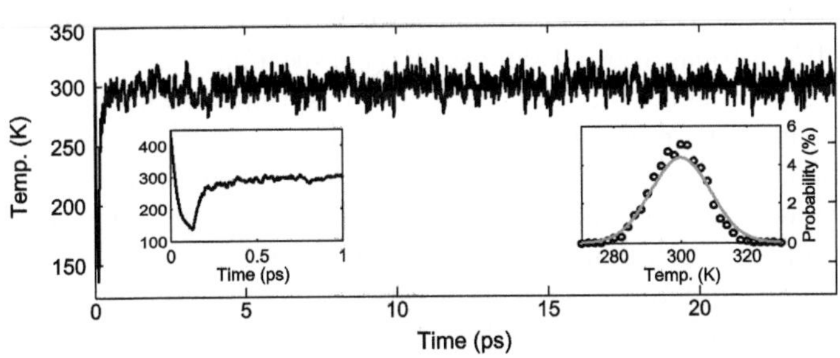

Fig. 7.10 The solvent temperature of the Ru=Co complex in the electronic ground state, showing the initial equillibration of the temperature. The first 130 fs are simulated using 0.25 fs steps at 50 K, with a friction of 0.1 on the entire system. The friction is set to 0.05 on the solvent only, for the production run-part. The *left inset* shows a zoom of the first ps. The *right inset* shows the actual versus theoretical temperature distribution of an NVT ensemble of finite size, as discussed in Sect. 4.3

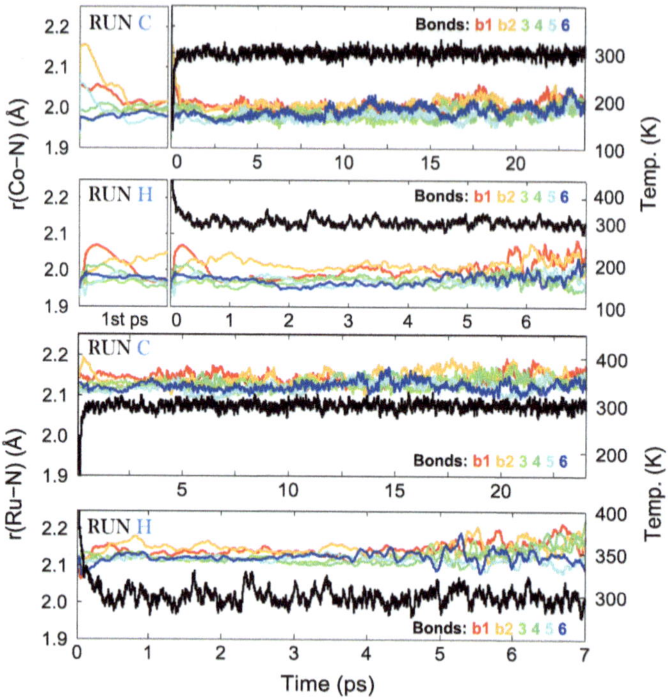

Fig. 7.11 Plots of the equillibration of the main structural parameters in the system, the Metal-N distances. The *top* two plots show the 6 Co–N distances for run C and H. The *bottom* two plots are for Ru–N. The distances have been smoothed with a 0.5 ps moving average to reveal any possible change on longer time scales. The Co–N dynamics in the first ps are shown in the inset in the *top* two plots, to more clearly reveal the initial equilibration process

of the metal-N bond distances of both run C and H, which takes place within the first few ps of the simulations.

For run H, the mean solvated structure is sampled at times >2 ps from the start of the run. The solvated mean, unconstrained structure of the run is shown in Fig. 7.12, using the 50 % thermal ellipsoid representation. Inspecting the representation reveals that the ellipsoid volumes (i.e. the nuclear disorders) increase when going away from the metal centers. Adapting the interpretation from [4], this is due to the increase in thermal fluctuations with increased exposure to the solvent. This observation, however, is contrasted with the much smaller ellipsoids in the bridge -region, underscoring the rigidity of the π-conjugated electronic configuration of the bridge.

The maximum difference of the mean Me–N bond lengths between runs C and H is 0.01 Å or 0.6 % of the bond length itself. Thus, For all other parts of the Ru=Co analysis, and in all other sections, we define the start of the production GS run C as after 6 ps, and use this trajectory. The two excited state spin manifolds LS and HS where started 4 ps into the GS C run.

The LS and HS runs were sampled for 10.5 ps and 21 ps, respectively.

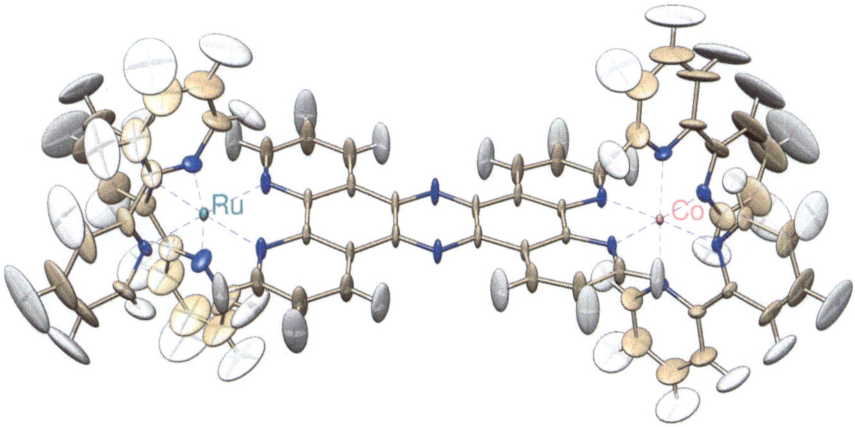

Fig. 7.12 50 % Probability- or thermal ellipsoid representation of the GS average structure of Ru=Co, made from run H The representation is made by aligning the Ru=Co geometry to the first frame, to remove rotations and translations of the entire molecule, using the RMSD Trajectory tool in VMD [5], and then exported to a new trajectory of aligned frames, from where the covariance matrix was calculated, exported in the 'anisou'-fields in a pdb file, and visualised using Chimera [6]

7.3 Thermal Distributions

7.3.1 Bond Lengths

Figure 7.13 shows the effect on the Metal-N bonds of changing the spin manifolds from the GS configuration, to the LS and HS, respectively, i.e. employing the same approximation for electronic excitation as in the case of $[Ir_2(dimen)_4]^{2+}$. In contrast to the $[Ir_2(dimen)_4]^{2+}$ system, the different spin configurations do not necessarily share similar PES-shapes. Furthermore, it must be remembered that the simulated GS→HS transition is artificial, which again must mean that the cooldown-time cannot be expected to be comparable to experimental results. Therefore, we limit ourselves to what one could call 'steady-state' sampling of thermal distributions of the cooled structures. Thus, leaving the thermostat on the solvent part of the system does not pose any problems to the results.

In the LS state, the Co–N bond vibrations seem to increase in amplitude, but no signs of systematic dissipation of vibrational energy is seen to occur within the 10 ps sampled. Bond lengths b1, b2, 4, and 5 seem to increase, while 3 and 6, the two bonds out of the bridge-plane, keep the same average length.

For the HS trajectory, we observe a clear increase in all bond lengths, accompanied by systematic vibrational dissipation of excess energy, as expressed in the decreasing amplitude in oscillations of both Ru–N and Co–N bonds, lasting approximately 6 ps. The final state having an overall on average much smaller vibrational amplitude than the other two states. For this state, we sample the distributions at times $t > 6$ ps. For

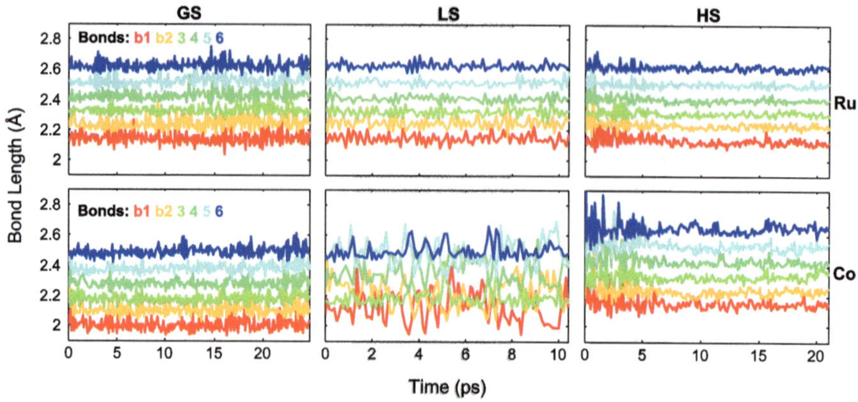

Fig. 7.13 Evolution of the Metal-N bonds in Ru=Co. The data has been re-binned in 10 fs averages, and each distance after the first is raised with 0.1 Å more than the previous distance, so that the individual trajectories can be seen. No vibrational cooling is observed for the LS excitation, but for the HS, the vibrational motion in both metal centres is seen to cool down in roughly 6 ps

the LS state, no vibrational cooldown is observed within the 10 ps simulation, even though the overall structural changes are smaller in this case. Therefore, the thermal distributions of the relevant parameters are sampled from $t > 2$ ps.

The thermal distribution of the Ru–N and Co–B bonds are visualised in the first and second rows of Fig. 7.14, respectively. The first two plots in each row represent the bonds formed with the nitrogens in the bridge. The average Ru–N bond lengths are almost unchanged by the excitations, but the HS distribution becomes almost 5 times as narrow as in the GS. This is also the case for the Co–N bonds, in contrast to the *widening* of the Fe–N bond distributions of $[Fe(bpy)_3]^{2+}$ [4]. The Co–N bonds are elongating in the HS state, with an average of 0.14 Å, 0.06 Å less than the experimental result [III, V]. The cause for this discrepancy is analysed in Sect. 7.4, but first we continue with the thermal distributions.

The apparent splitting of the bond length in some cases for the LS Co–N bonds are further exemplified in Fig. 7.15. The splitting is not evident in the bonds to the bridge-nitrogens, but takes place in various amounts in the rest of the bonds, and example of which is shown in the right plot. For the non-bridge bonds, the distribution is clearly not equivalent to that of a single gaussian. Since we do not observe any constantly increasing or decreasing bond lengths in Fig. 7.13, we can rule out that this the splitting is caused by the system not having equilibrated fully. The smaller sub-distribution has bond lengths more comparable to the HS state.

7.3.2 Bite Angles

For the excited state, solvatisation does seem to have an effect on some of the bite angles, as observed in Fig. 7.16: In LS, solvatisation reduces the angles, whereas in

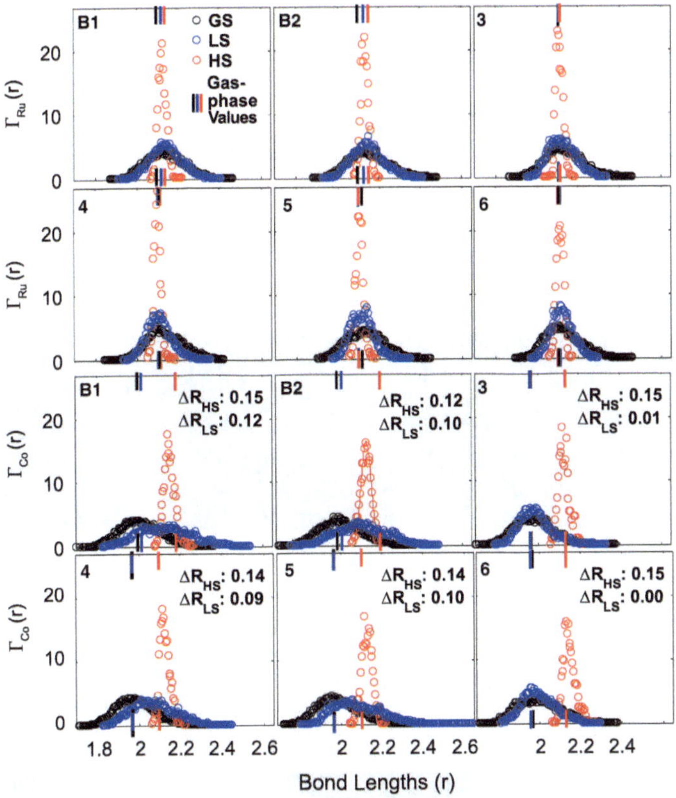

Fig. 7.14 Thermal Distributions of the Ru–N (*top row*) and Co–N (*bottom row*) bonds. The x-axis marks represent the vacuum values in the same colour code as the thermal distributions. ΔR is the average bond length change of each Co–N bond, w.r.t the average GS bond length of each bond. In the Ru-centre, almost no change in average bond length from GS to the excited states are seen, the maximum being a 0.03 Å contraction in the HS state, for two of the bonds. The gas-phase lengths are very similar to the solvated averages. The six HS distributions are much narrower, with an average standard deviation of 0.02 Å, compared to 0.09 Å and 0.06 Å for GS and LS, respectively. For the Co–N bonds, an average lengthening of 0.14 Å is observed in the HS state. In the LS state, some bond elongation seems to occur, and the thermal distribution becomes even wider, with indications of double peaks or shoulders. Finally, the LS solvated average bond lengths are longer than the vacuum values

the HS state, one angle is reduced, one is enlargened, and one does almost not change. The figure also reveals that all Co-centre bite angles of the HS state are reduced by approximately 5°, and the distribution becomes narrower. The bite angles of the Ru-centre are unaffected within the bin width, and are thus not shown.

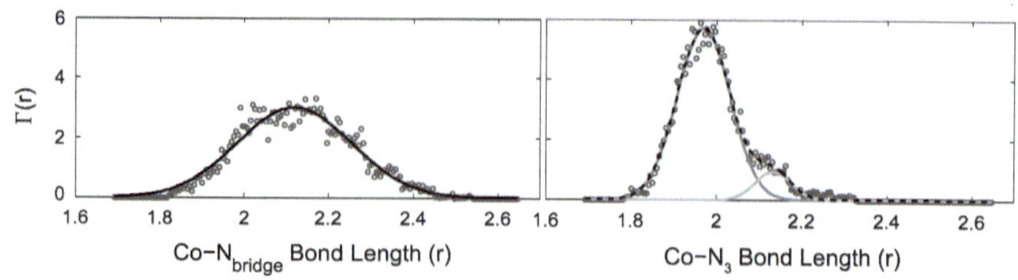

Fig. 7.15 LS Thermal distributions of a Co–N_{B1} bond (*left*), and Co–N_3 (*right*). The distribution of the non-bridge bonds split up, where one part has bond lengths more comparable to the HS state

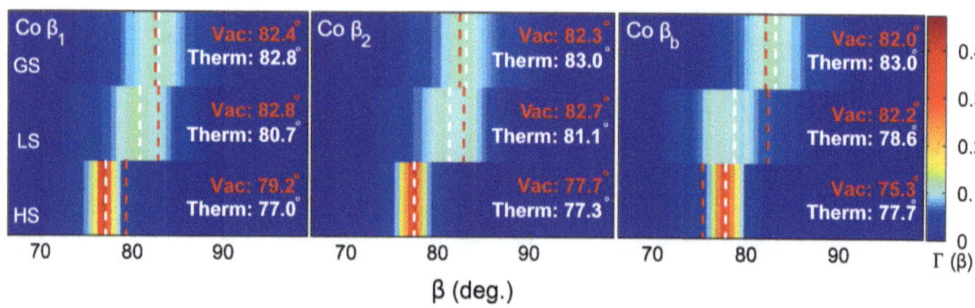

Fig. 7.16 The Ru=Co bite angles of the Co-centre, sampled in bins of 0.5°. The HS distribution becomes narrower, similar to the bond length distribution, and all three bite-angles decrease ~5° in the HS state. The Ru–N centre results are not shown, since the angles remain largely unchanged. The *dashed red lines* show the gas phase results, while the *dashed white lines* show the average of the solvated distribution

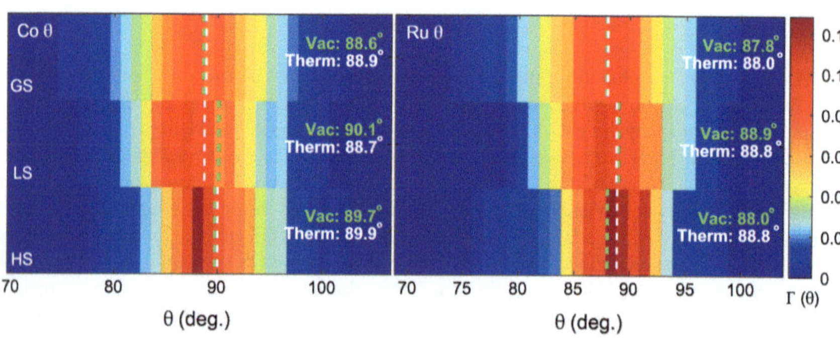

Fig. 7.17 The butterfly angles θ of both the Ru- and Co-centre. The angles remain largely unchanged by both excitation and/or solvation

7.3.3 *Butterfly Angles*

Figure 7.17 shows the θ-angles of both centres. The average values are generally affected by neither solvation nor excitation, but are in general 'softer' (i.e. more widely distributed) than the bites, and no change of this occurs in the excited states.

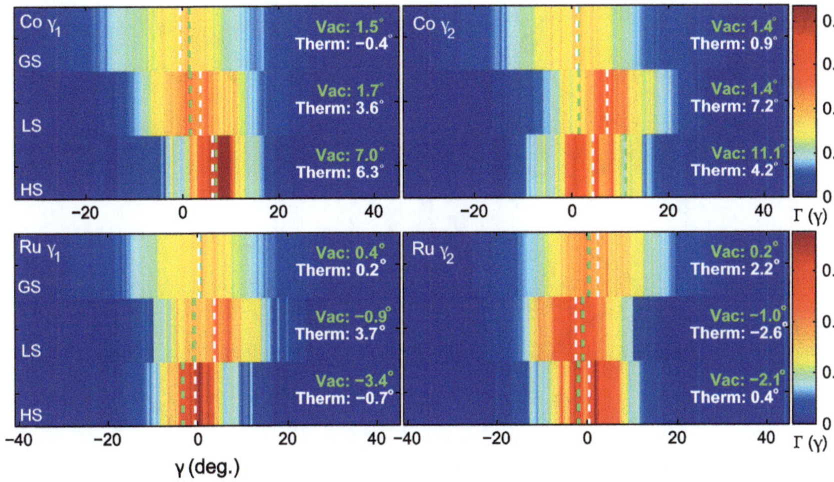

Fig. 7.18 The torsion angles for the Co-centre (*top*) and the Ru-centre (*bottom*), binned in 0.5° bins. The excited state angles for the Co-centre are slightly larger than for the ground state. For the HS Ru-centre, the solvation seems to bring the angle back to the GS values

7.3.4 Torsions

The Co-centre torsion angles shown in Fig. 7.18 are enlargened by the HS excitation, similar to the $[Fe(bpy)_3]^{2+}$ case [4]. Here, we also show the Ru-centre results, since we observe that the small change in the gas-phase HS torsion angle seem to be counteracted by the solvent. In the Co-centre, the solvent similarly reduces the torsional twists in the complex.

7.4 Ru=Co Orbitals

In order to rationalise the shorter simulated bond elongation compared to the experimental value of 0.20 [V] Å, Fig. 7.19 visualises the orbitals around the HOMO-LUMO levels of each of the simulated electronic states. In the GS, the LUMO+1 orbital visually resembles the HS_α HOMO, and the GS LUMO+2 is a bridge-localised orbital which could be speculated to play a role in the real-world excitation relaxation cascade [III]. The LS_α HOMO resembles the GS LUMO.

We now turn our attention to the β-spin channel of the spin-unrestricted simulations. As already discovered during the gas-phase optimizations, reaching both structural- but also even just wave-function- convergence was challenging. These problems persisted for the actual QM/MM MD runs, which ment it was necessary to maintain the Fermi-Dirac-smearing of occupation numbers, the result of which is seen in the figure inset. The occupation of the orbitals in the β-channel is effectively

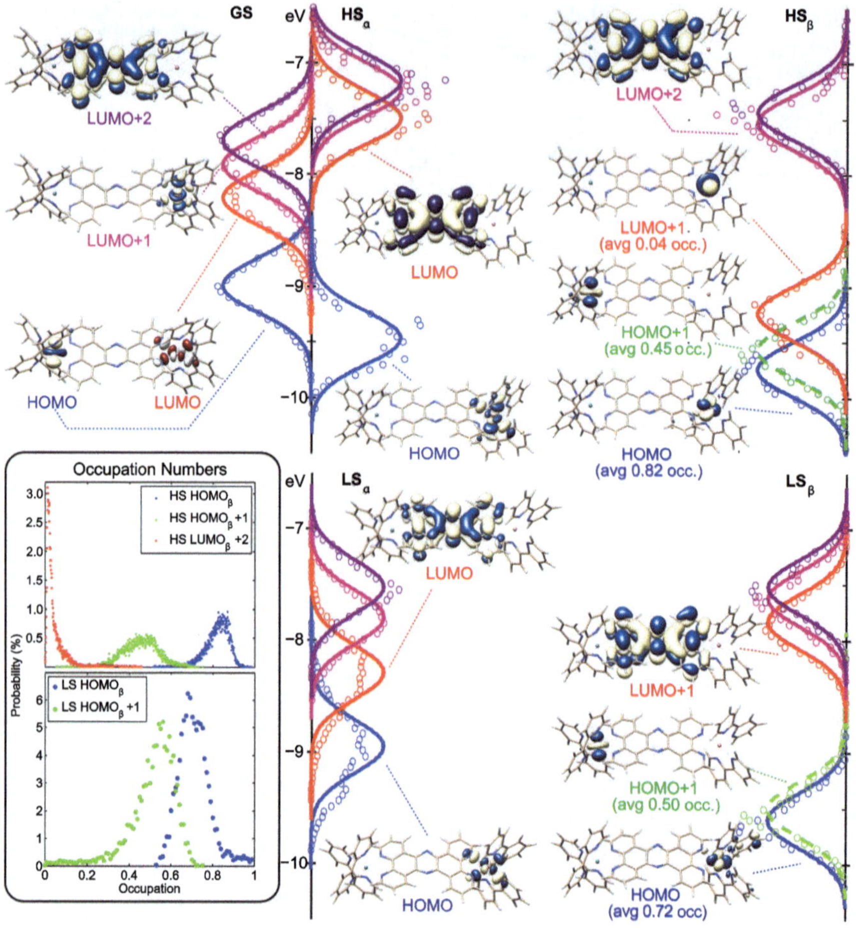

Fig. 7.19 Overview of the relevent occupied and virtual orbitals for Ru=Co, in the ground state, and the two ES spin manifolds of LS and HS. We observe that the GS LUMO is similar to the LS and HS α spin HOMO. The energy levels are sampled from each time step, and are seen to be distributed with a variance of 10 $k_b T$. Nothing has been done to improve on the absolute energies of the virtual orbitals, and as such they are mostly displayed for completeness. The ES β-spin channels suffer from substantial charge-smearing, as shown in the lower left inset. This is due to the finite Fermi-Dirac width which had to be applied for converging the SCF-optimizations within a feasible amount of steps for the QM/MM MD simulations

distributed such that the LUMO on average has an occupation number of almost 0.5, which is why it in the figure is labeled ‘HOMO+1’ instead. From the depicted isosurfaces of this orbital, we see that it is localised on the Ru-centre, meaning that not all charge has been successfully transferred to the Co-centre localised orbital, and therefore the resulting Co–N bond elongation is smaller than the experimental result.

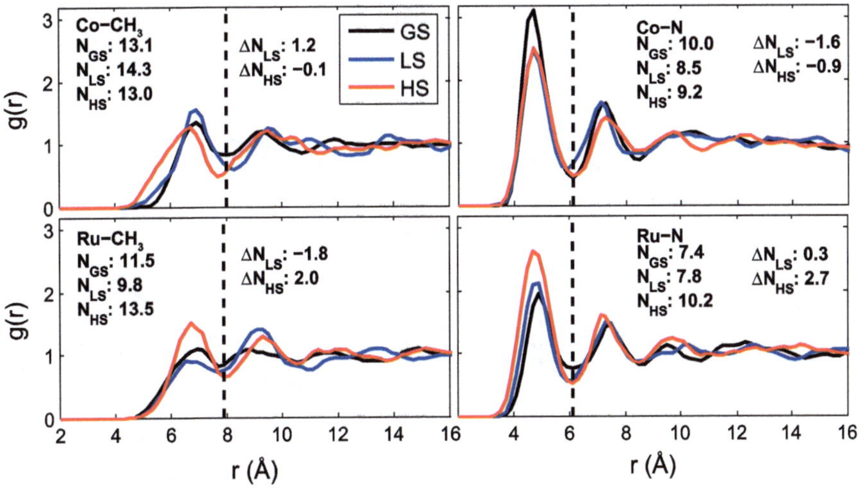

Fig. 7.20 Pairwise RDF between each of the metals and each end of the solvent molecules. The *dashed lines* indicate the chosen solvent shell definitions. The RDF are sampled from the equilibrated parts of the three trajectories, with a Δr of 0.2 Å and are shown here after a 3 point moving average smoothing. The N-values are coordination numbers of each state, and ΔN are the differences from the GS to the two excited states. The Ru-centre increases its coordination number in the HS state

This distribution of occupation in the levels will most likely also affect the solvent-response, but seeing as how we do indeed model part of the charge transfer, and the resulting bond-elongation, we now analyse the effects of the excited-state charge transfer in Ru=Co on the solvent.

7.5 Solvent Interaction

Figure 7.20 displays the RDF between the metal centres and each end of the ACN solvent.[2] The dashed lines demarkate the solvent shell definitions, and we observe that the first N peak is found on average roughly 2 Å closer to the metal than the CH_3-peak, with the ACN-molecule having a length of 2.62 Å. In the excited states, the N-peak is slightly shifted towards Ru, which can be attributed to the partially negatively charged N-ends on ACN responding to the Ru oxidation. The Co–N peak is dampened by the excitations. The figure also displays the coordination numbers, which for the Ru-centre is lower than the literature result for $[Ru(bpy)_3]^{2+}$ [4], which is expected, since the bridge must sterically hinder the solvent, to a certain extent. Finally, the changes in coordination numbers when going from the ground state to the excited states are also shown. In the HS state, the coordination number increases to

[2] While the abbreviation 'Me' was used for the **me**thyl group in the previous sections where systems had only one type of metal atoms, we here explicitly denominate the methyl group 'CH_3', to avoid confusion with the umbrella-term 'Me' here used for both Ru and Co **Me**tals.

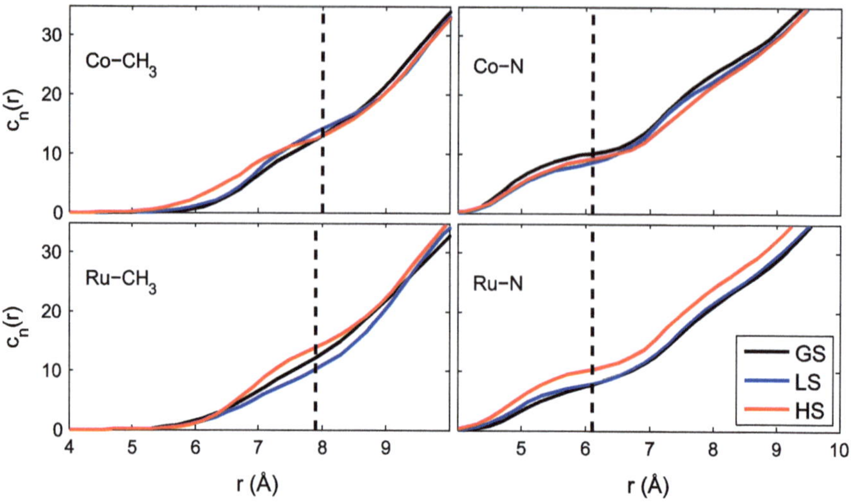

Fig. 7.21 Cumulative coordination numbers, $c_n(r)$ of the Metal-Solvent correlations

the extent of 2–3 ACN. We note in passing that this is an example of a specific solvent response, where the Ru-centre shows the opposite behaviour of what is believed to take place for $[Fe(bpy)_3]^{2+}$, as described in Chap. 4, an effect which could not be reproduced with the more simple, purely classical MM model.

Returning to Ru, A similar effect has been observed in the case of water-solvated $[Ru(bpy)_3]^{2+}$, where Ru(III) is shown to accommodate one water molecule more than Ru(II) [7]. The Co-centre experiences a 0–1 molecule decrease in the HS state, meaning that the solvent molecules are not simply transferred from the Co end to the Ru end. The overall charge of the complex is 5+, so the Co-centre is still N-coordinating, even after the excited state reduction. At the time of writing, the LS state suffers from only being sampled for 6.5 ps, which is roughly 1/3 of the GS and under half of the HS sampling time, and thus giving less reliable distributions.

The plots in Fig. 7.21 display the cumulative coordination numbers $c_n(r)$ from the metal centres, to the solvent. The slope is lower in the regions where the bpy ligands are located, and for the Co–N curve, there is an overall decrease in coordination for any $r < 10$ Å in the two excited states, whereas the opposite is the case for the Ru–N curve.

Figure 7.22 shows the N:CH_3 cumulative coordination number-ratio for each of the metal centres. As expected, the ratios converge towards one as the distance from the solute increases. At closer ranges, the ratio increases for both centres, indicating an overall preferred orientation of pointing the ACN N-end towards the metals, as expected. The Co-ratio decreases upon excitation, indicating that, on average, the N-ends become less oriented towards Co compared to ground state, while still maintaining some of the preferred N-Co orientation. This is consistent with the Co-centre being reduced in the excited states, which affects the partially negatively charged N-ends of ACN accordingly. Likewise, but to the opposite effect, the ratio increases for the Ru-centre.

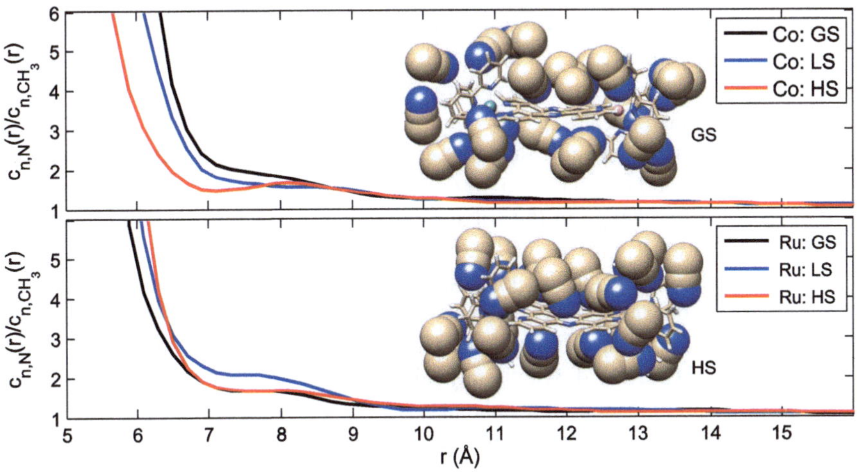

Fig. 7.22 The N:CH3 ratio of the cumulative coordination numbers for Co (*top*), and Ru (*bottom*). The two images depict snapshots of the GS (top) and HS (*bottom*) solvent shells, defined by drawing only the full solvent molecules which are within 8 Å of either Ru (*left on the images*) or Co. The ratios converge to 1 at long distances, as expected

Even though the full charge transfer is encumbered by the Fermi-Dirac smearing of the electronic occupations, the solvent responds in a manner which can be rationalised as done here. The solvent shell changes observed around the Ru-centre will most likely play a role in the upcoming analysis of data collected at XFEL beamtimes, since it will correspond to a local change in the density.

7.6 Further Discussion and Conclusions

For the computationally most demanding of the complexes in this work, we have so far been able to set the scene for later, more in depth studies of the electron transfer process. Despite the GPAW issues with convergence in the excited states, the employed model qualitatively reproduces the experimental bond-length elongation in the Co-centre in the HS state, and the quantitative deviations can be explained by looking into the occupation of the relevant orbitals.

A further study of the geometric changes originating from the Co–N expansion revealed that the HS bite-angles are reduced, while the Co-centered torsions of the excited states are slightly enlarged, all of which is similar to what has been reported for $[Fe(bpy)_3]^{2+}$ [4]. The only difference is in how the width of the distributions respond to the excitation, where we observe that all Metal-N bonds become much less floppy in the HS state. Since neither this (yet), nor the $[Fe(bpy)_3]^{2+}$ study has the luxury of basing interpretations on a statistically more sound foundation of configuration sampling,[3] it is still too early to comment further on this discrepancy. Recalling the previous chapter, if individual trajectories were singled out in the study of $[Ir_2(dimen)_4]^{2+}$, both extremes in the conclusions about the floppiness of the ES Ir-Ir bond could have been reached.

The findings on both specific- and non-specific solvation dynamics are interesting in their own right, and in relation to the conclusions of Chap. 4. For Ru=Co, we have observed a specific change in the solvent-intercalation, akin to how $[Ru(bpy)_3]^{2+}$ is believed to interact with water [7]. Since we have also seen that the geometry of the Ru-centre remains largely unchanged by the excitation, the specific solvation dynamics are in this case mostly caused by the electrostatic changes experienced by the complex, which again explains why these processes are not reproduced by the MM MD method of Chap. 4, which only takes the geometric changes into account.

Since the obtainable amount of solvent configuration sampling is computationally limited when employing QM/MM methods, one could think of a future compromise where better statistics are needed for the experimental comparisons: By taking the averaged structure(s) from shorter, equillibrated QM/MM MD runs (e.g. as depicted in Fig. 7.12), *and* using the associated, averaged partial charges on the atoms in MM MD simulations with fixed solute geometries, we hope to sufficiently improve the MM approximation to reproduce realistic solvent shells, without having to resort to the computationally much more costly parallel/longer QM/MM sampling. The Direct Dynamics can then be reserved for out-of-equilibrium dynamics, etc.

All in all, even while the Ru=Co system seems to be pushing the limits of the GPAW capabilities, the QM/MM MD method is again able to provide insights into structural changes and the solvent interplay in the system.

References

1. A.D. Becke, Phys. Rev. A **38**, 3098 (1988)
2. J.P. Perdew, Phys. Rev. B **33**, 8822 (1986)
3. Y. Zhao, J. Pu, B.J. Lynch, D.G. Truhlar, Phys. Chem. Chem. Phys. **6**, 673 (2004)
4. L.M.L. Daku, A. Hauser, J. Phys. Chem. Lett. **1**, 1830 (2010)
5. W. Humphrey, A. Dalke, K. Schulten, J. Mol. Graph. **14**, 33 (1996)
6. C. Huang, G. Couch, E. Pettersen, T. Ferrin, *Chimera: An Extensible Molecular Modeling Application Constructed Using Standard Components* (Pacific Symposium on Biocomputing, 1996). http://www.cgl.ucsf.edu/chimera
7. J.J. Szymczak, F.D. Hofmann, M. Muewly, Phys. Chem. Chem. Phys. **15**, 6268 (2013)

[3]e.g. Through the initiation of more parallel ES trajectories.

Part IV
Summary

Chapter 8
Summarising Discussion and Outlook

The investigation of transient dynamics in transition metal complexes and their surroundings through simulations has led to the observations presented in this thesis. The following is an attempt to condense the discoveries, discuss future strategies, and return to the initially posed questions on how excess excitation energy dissipates in solvated systems.

First, we needed to show that the answers we have produced would be trustworthy. Thus, the first part of the project was concerned with ensuring that the theoretical modelling of the transition metal complexes did not suffer from neglecting relativistic bond-length distortions. Therefore, in Chap. 3 we benchmarked the most common relativistic approximations to geometries derived from using the fully relativistic Dirac-Coloumb Hamiltonian on the $[Pt(CN)_4]^{2-}$ complex, and compared the bond length changes to changes induced by varying other approximations, such as the exchange-correlation functional used. By employing the Dirac-Coloumb Hamiltonian, we found a contraction of the Pt–C bond, ascribed to the relativistic relaxation of the kinetic repulsion. We observed that the relativistic effects were sufficiently describable within the GPAW-framework, which possesses the required computational efficiency for the later Direct Dynamics simulations.

The next (Chap. 4) was concerned with how to calculate an experimental XDS signal from molecular simulations, for comparison to experiment. We used simple MM MD simulations to approximate the signal from the (changes in the) solvent-shell around frozen molecular geometries of $[Fe(bpy)_3]^{2+}$, acquired from studies in literature using more advanced models. This 'first order approximation' could successfully be used in a fit of experimental data.

We then addressed the technical issues with simply employing the most common Debye-formulation by deriving an expression for the scattering based on the RDF, and confirming its numerical implementation. The physical issues of the 'first order approximation' were then analysed by comparison of the solute-solvent interactions through solute-solvent RDF_S. We observed how, in this case, classical force-field MD methods can capture parts of the overall, non-specific features of solvation, and can thus be used in experimental fits, but we also observed how they do not necessarily capture the full solvent response, including the possibility of specific

A.O. Dohn, *Transient Changes in Molecular Geometries and How to Model Them*, Springer Theses, DOI 10.1007/978-3-319-18747-1_8

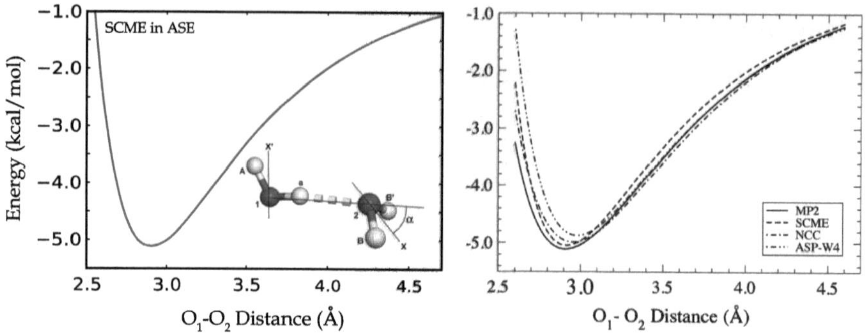

Fig. 8.1 The first step of implementing a polarisable force-field in ASE. *Left* O–O Dimer PES, from the ASE-interfacing of the original force field code. *Right* Results from the original force-field paper [1]

solvation dynamics. We argued that including electronic structure changes in this model can only be done in a somewhat approximative, heuristic fashion.

Chapter 5 introduced our solution to these issues, providing background and benchmarks on the Direct Dynamics method, and how this method functions in our implementation. Since many of the experiments carried out in our group are conducted in acetonitrile, such a force-field was implemented, tested, and used in the discussion about the QM-MM interfacing region. The overall pseudo-structure of the ACN solvent is well reproduced in the QM/MM model of a single QM ACN molecule in a MM ACN 'solvent'. There were few discrepancies, which we demonstrated to originate from the unification of the QM and MM regions. In the current formulation, the intramolecular charge distribution on the MM particles, represented by the partial charges on the individual particles, remains fixed.

This is an issue which we hope to improve upon in the future, through employing polarisable force fields, which improves on the classical modelling of partial charges. For a brief outlook on how to achieve such improvements, we respectfully break the unwritten rule of not presenting new information in final sections,[1] to show the very first steps towards incorporating a polarisable force field for water [1]. The steps were taken during a one-month visit at the University of Iceland, and are visualised by the reproduction of the potential energy curve of the water dimer (Fig. 8.1). The left curve is calculated with an implementation of the polarisable force-field as a calculator object in ASE, showing that the original Fortran code has been successfully interfaced to the ASE environment in which our QM/MM implementation works. There is still much work needed on the actual QM/MM interfacing, but the first steps towards improving the description in the interfacing region have already been taken.

Attention was then turned to answering the question of how energy dissipates from a complex to the solvent. Here, the $[Ir_2(dimen)_4]^{2+}$ complex was used in the study of transient dynamic processes behind the energy dissipation. By employing

[1]Which, coincidentally results in this unwritten rule now being …written. In any case, the choice of shying away from labelling this section 'Conclusion' is deliberate.

our QM/MM implementation for Direct Dynamics MD simulations, we were able to reproduce the experimentally observed metal pinching oscillation, thus demonstrating that the implementation is fully capable of simulating the out-of-equilibrium IVR and EVR processes. Consequently, we could improve on the experimentally obtained model by analysing the pinch, twist, and breathing modes in relation to the solvent, and found that the rigidity of the ligands defined the spread of the population of GS-geometries, which again was the main influence for the loss of coherent motion. In this regard, the system of solvated $[Ir_2(dimen)_4]^{2+}$ is an interesting case, where the coherence lifetime is actually increased by solvation, since the solvent can block the IVR which would cause decoherence. The role of the solvent was further studied through the temporal evolution solute-solvent pairwise RDF_s, where we could show that even though the direction of the dipole moment does not change as a result of the excitation, the solvent *still* shows signs of reorganisation to stabilise the new excited state electronic configuration.

The final project employed the Direct Dynamics method to the bi-centered Ru=Co complex, in some ways an even more intricate system. Contrasted to $[Ir_2(dimen)_4]^{2+}$, it does not have direct metal-metal interactions, and it is believed to have a narrower GS thermal distribution of geometries. However, the structural changes in the complex due to excitation are more subtle.

The goal here was to study the effect of solvation on an intermediate- and the final spin-state of the excitation relaxation cascade. Our method was able to qualitatively reproduce the Co–N bond elongation, and the cause for the quantitative discrepancy could be explained by the technical necessity of having to apply a finite Fermi-Dirac distribution of the occupation of orbitals close to the HOMO-LUMO gap, in order to reach SCF convergence. A future correction of this problem could involve *constrained DFT*, a method which employs an external potential to further constrain the resulting electronic distribution, which has been used successfully before [2, 3].

Despite this issue, we were able to observe both specific- and non-specific solvation dynamics of the solvent shell around the complex, as a consequence of the excited state electron transfer. The Ru-centre solvent is able to intercalate more solvent molecules when excited, similar to what has previously been observed for $[Ru(bpy)_3]^{2+}$ in water [4], and moving the charge from Ru to Co affects the overall orientation of the ACN_s accordingly. Thus, we see that, by using our QM/MM MD implementation, we *recover* the processes we could *not* describe by employing purely classical force-field methods of Chap. 4.

Future scenarios could easily arise, where employing the Direct Dynamics strategy for a statistically exhaustive amount of sampling will not be justifiable, regarding the amount of information obtainable from such a non-trivial set of simulations. In this case, we propose to combine the 'frozen-solute'-approximation from Chap. 4 with the Direct Dynamics method. This would be done by obtaining averaged molecular geometries and partial charges through a shorter QM/MM MD simulation. Then these parameters should be used for the frozen solute, in a longer MM MD solvation sampling simulation. This should amount to an increase in precision, compared to approximations using frozen solutes with partial charges obtained from ab initio simulations of complexes in gas-phase.

We are currently limited in the QM description to ground state DFT, and approximating electronic excitations by changing spin-states. One option for simulating excited states with the same multiplicity could be the ΔSCF method [5, 6], which is already implemented in GPAW. Interfacing the ΔSCF-calculator with the MM calculator could be a possibly fruitful project in the future.

The current limitations in our Direct Dynamics method is overshadowed by its efficacy, and this thesis will hopefully represent itself as a testimony to its promise. While many QM/MM MD studies are still confined to interpreting results from single trajectories of few picoseconds, this implementation has proved its capabilities in terms of efficiency, accuracy, and the power to explore the world of femtochemistry.

References

1. K.T. Wikfeldt, E.R. Batista, F.D. Vila, H. Jónsson, Phys. Chem. Chem. Phys. **15**, 16542 (2013)
2. Q. Wu, V. Voorhis, Phys. Rev. A **72**, 024502 (2005)
3. H. Oberhofer, J. Blumberger, J. Chem. Phys. **131**, 064101 (2009)
4. J.J. Szymczak, F.D. Hofmann, M. Muewly, Phys. Chem. Chem. Phys. **15**, 6268 (2013)
5. J. Gavnholdt, T. Olsen, M. engelund, J. Schiøtz, Phys. Rev. B **78**, 075441 (2008)
6. T. Olsen, J. Gavnholt, J. Schiøtz, Phys. Rev. B **79**, 035403 (2009)

Appendix A
Ru=Co Gas Phase Geometry Optimization

Here, the rest of the figures used in obtaining the gas-phase structural parameters for Ru=Co shown in Table 7.1 are collected.

A.O. Dohn, *Transient Changes in Molecular Geometries and How to Model Them*, Springer Theses, DOI 10.1007/978-3-319-18747-1

Torsions

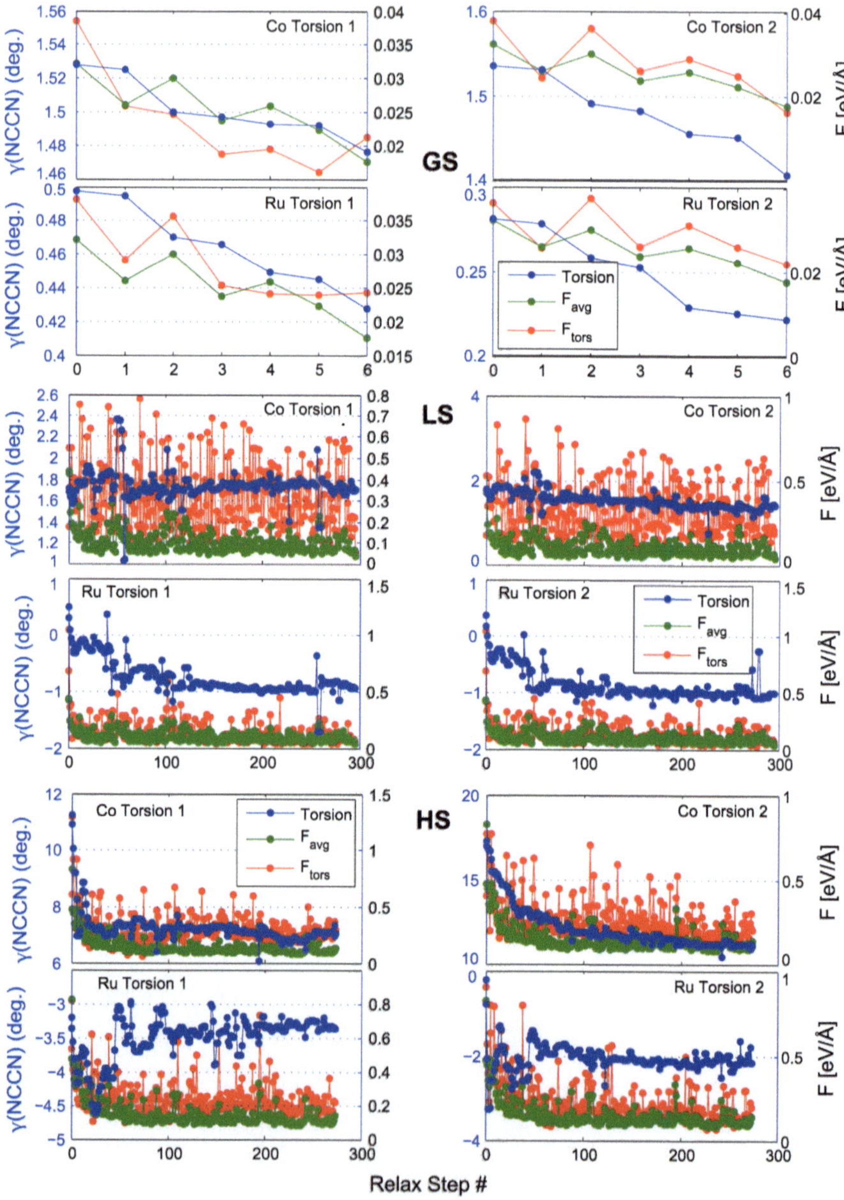

Fig. A.1 Vacuum relaxed torsion angles, γ

Butterfly Angles

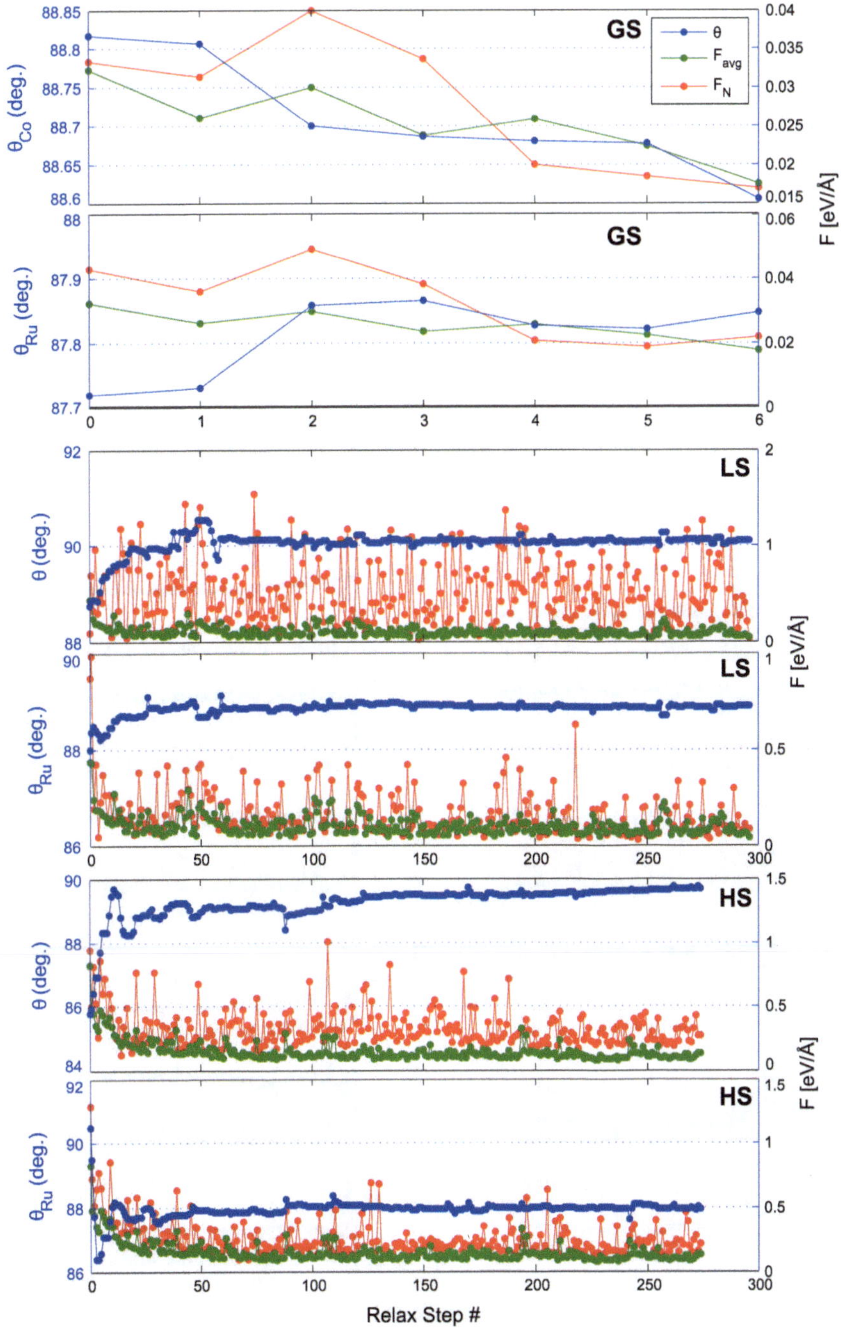

Fig. A.2 Vacuum relaxed butterfly angles angles, θ

Bite Angles

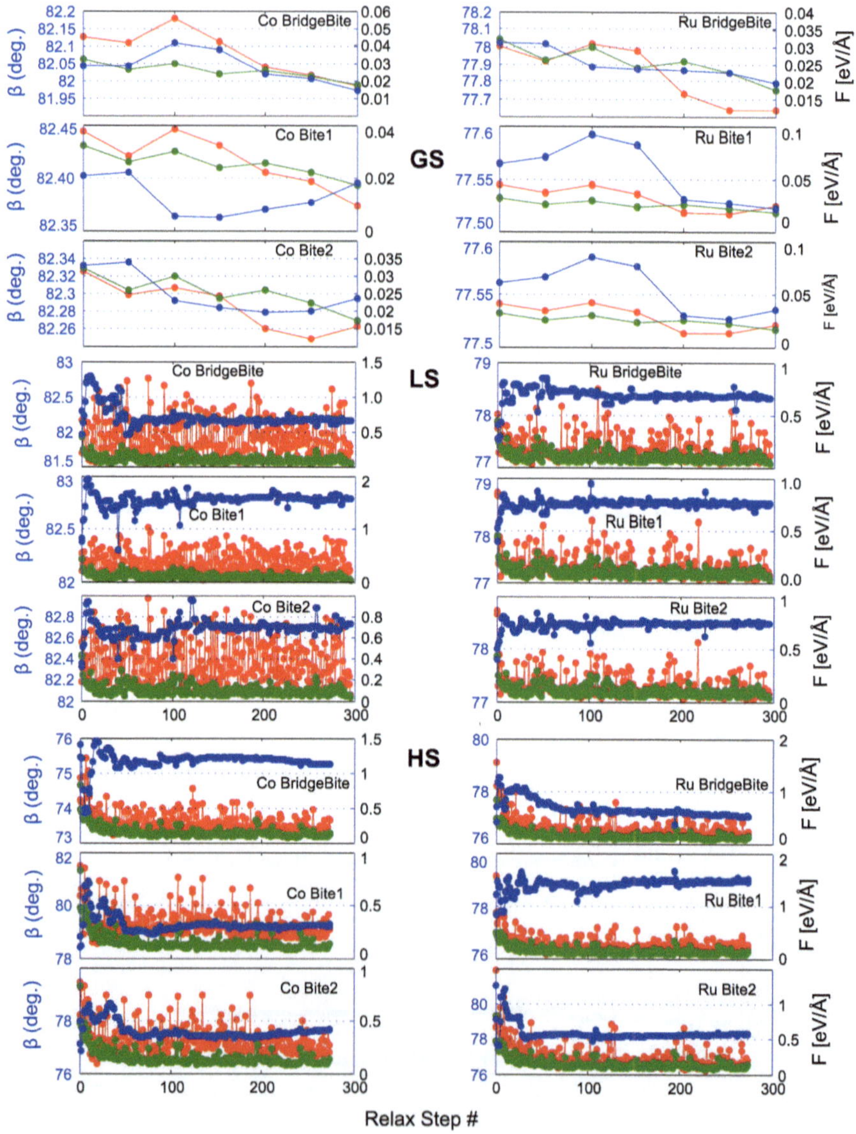

Fig. A.3 Vacuum relaxed bite angles, β

Appendix B
Scripts of Relevancy to Future Users

In the hope that some of the developed scripts may find use with people working with similar problems, the most widely useful are presented below. It should be evident that each script in no manner represents the only way to solve the posed problem, but can serve as a template or inspiration, e.g. for future students, which may have difficulties finding help with these problems online or in textbooks, due to the somewhat unique combination of specificity and generality of day-to-day problems in scientific computation. While USB sticks, CDs or Dropbox links are convenient, they have a tendency to vanish, whereas paper can be firmly placed upon a shelf, and be taken out as a final resort …

B.1 ASE/GPAW Scripts

A.O. Dohn, *Transient Changes in Molecular Geometries and How to Model Them*, Springer Theses, DOI 10.1007/978-3-319-18747-1

Listing 1: Solvent box creation script, originally made for water, by E. Ö. Jónsson

```
from ase import *
from ase.io import read, write
from ase.visualize import view
from math import pi
import numpy as np

# Original by E. O Jonsson
# I basically just switched water to ACN
# And made the code a bit more "stupid"
# W.R.T pre-randomization of the box since
# ACN is linear so it's a bit harder to
# randomly orient. And it's easer just
# to get ASE to do it. /Asmus Dohn 2012

# Make MM aceto
aceto = Atoms('C2N')
pos = np.array([[0, 0, 0],      # distances from article
                [0, 0, 1.458],
                [0, 0, 1.458+1.157]])

for a in [0,1,2]:
    aceto[a].set_position(pos[a])


atomno = len(aceto)

write('new.xyz', aceto)

aceto.set_cell(([32.,0,0],
               [0,32.,0],
               [0,0,44.]))

cell = aceto.get_cell()
vol = cell[0,0]*cell[1,1]*cell[2,2]

# Total number of aceto
no_aceto = round(vol*0.01153)
# Number of layers along z-axis
no_l = cell[2,2]/3
# Number of aceto in each layer
no_aceto_l = round(no_aceto)/no_l
# Number of aceto along x and y direction
no_aceto_xy = round(np.sqrt(no_aceto_l))

print "no_l: "+ str(no_l)
print "no_aceto_l: "+ str(no_aceto_l)
print "no_aceto_xy: "+str(no_aceto_xy)

#arr = np.zeros([round(no_aceto_xy)**2,2])
```

```
# Create coordinates
arr = []
for i in range(int(no_l)):
    for j in range(int(no_aceto_xy)):
        for k in range(int(no_aceto_xy)+1):
            arr.append([k*cell[0,0]/(no_aceto_xy+1),j*cell[1,1]/no_aceto_xy, i*
                cell[2,2]/no_l])

print "no_aceto: "+str(no_aceto)
#print "real number: "+str(np.shape(arr)[0])

l = 0
for k in range(np.shape(arr)[0]):
    new = read('new.xyz')
    for i in range(atomno): new[i].x += arr[k][0]
    for j in range(atomno): new[j].y += arr[k][1]
    for m in range(atomno): new[m].z += arr[k][2]
    if k%2==0:
        new.rotate('x', pi/2., center=(new[1].position))
    else:
        new.rotate('x', -pi/2., center=(new[1].position))
    new.rotate('z', np.random.rand()*pi/3., center=(new[1].position))
    aceto += new

# displace
for j in range(0,len(aceto),6):
    aceto[j].x += 0.5
    aceto[j+1].x += 0.5
    aceto[j+2].x += 0.5

# displace
for j in range(0,len(aceto),6):
    aceto[j].y += 0.2
    aceto[j+1].y += 0.2
    aceto[j+2].y += 0.2

# displace
for j in range(3,len(aceto),6):
    aceto[j].y -= 0.5
    aceto[j+1].y -= 0.5
    aceto[j+2].y -= 0.5
# Delete the first aceto which is on top of #2
del aceto[0:3]

realno_aceto = len(aceto)/3.
diff = realno_aceto - no_aceto
```

```
print "# Of molecules difference from proper density: "+str(diff)

if diff > 0:
    delarr = np.zeros(realno_aceto)
    interv = round(realno_aceto / diff)
    delarr[0::interv] = 1
    delarr[1::interv] = 1
    delarr[2::interv] = 1
    delidx  = np.where(delarr == 1)
    delidx = list(delidx)
    delidx = np.flipud(delidx)
    for i in range(len(delidx)):
        del aceto[delidx[i]]

print "final # of solvent molecules: "+str(len(aceto)/3)

aceto.center()

# set Methyl masses for Me
masses = aceto.get_masses()
H3mass = sum(Atoms('H3').get_masses())
masses[::3] += H3mass
aceto.set_masses(masses)
test = aceto.get_masses()
print test

write('Box_MoreGridish.traj',aceto)
view(aceto)
```

Listing 2: QM/MM MD input script template, based on the ES runs of the Iridum dimer

```
### Import modules ###
from ase.io import read, write # for ASE I/O operations
from ase.visualize import view # visualization
from gpaw.calc_mm_inter import Inter_Pot # QM/MM Calculator
from gpaw import * # GPAW
import numpy as np
from gpaw.langevin_con_general import LangevinC as LC # MD
from ase.io.trajectory import PickleTrajectory as PT # Output module
import ase.units as unit # Units
from gpaw.ase_qmmm_TEST4 import *   # QM/MM Interfacer
# Tools
from gpaw.poisson import PoissonSolver as PS
from gpaw.occupations import FermiDirac as FD
from ase import Atoms
```

```
### Set I/O paths and names ###
PATH = '<Path to working directory>'
startno = 9 # Which d(Ir-Ir) exciation starts from
runno = 0   # Which run in the individual excitation
inname = 'IrD_ES_SSw_NF'+str(startno)+'_'+str(runno)
outname = 'IrD_ES_SSw_NF'+str(startno)+'_'+str(runno+1)

### Define QM Region ###
qm = read(PATH+'../../IrDimen_oShell.cube')
qmidx = len(qm) # The entire QM subsystem

### Constraining Hydrogens ###
qm_con =  # A list containing atomic indexes and distances of all hydrogens and
    the atoms they are connected to.
# Format example for a methyl group:
# [[(C1,H1,r1),(C1,H2,r2),(C1,H3,r3), (H1,H2,r4),(H1,H3,r5),(H2,H3,r6)], ... ]
# Where Cx and Hx are the atomic indexes, and r the distances, respectively.

### Read in already equilibrated system ###
qmmm = read(PATH+'<AlreadyEquilibratedSystem>.traj@-1')

### Set Charges ###
charges = np.zeros(len(qmmm))
charges[qmidx::3]   = .206
charges[qmidx+1::3] = .247
charges[qmidx+2::3] = .453
qmmm.set_charges(charges)

### Turn off GPAW-style PBCs - they are handled in ase_qmmm ###
qmmm.set_pbc(False)

### Create LJ array for the MM subsystem ###
# Dimensions: [(eps,sigma),atom index]
LJ_mm = np.zeros((2,len(qmmm[qmidx:])))
LJ_mm[0,0::3] += 0.7825 * units.kJ / units.mol
LJ_mm[0,1::3] += 0.544  * units.kJ / units.mol
LJ_mm[0,2::3] += 0.6276 * units.kJ / units.mol
LJ_mm[1,0::3] += 3.775
LJ_mm[1,1::3] += 3.650
LJ_mm[1,2::3] += 3.200
```

```
### LJ array for QM subsystem ###
# Assignment via ASE atoms.symbol string:
LJ_qm = np.zeros((2,qmidx))
for j,k,l in [['H', 6.81e-4, 1.487],
            ['N', 7.37e-3, 3.64],
            ['C', 4.74e-3, 1.908],
            ['Ir', 0.0190, 2.31]]:
    for i in range(qmidx):
        if qmmm[i].symbol == j:
            LJ_qm[0,i] = k
            LJ_qm[1,i] = l

### Change the ACN Me mass from C to Me ###
masses = qmmm.get_masses()
me_mass = sum(Atoms('H3C').get_masses())
masses[qmidx::3] = me_mass
qmmm.set_masses(masses)

### Fix magnetic moments to triplet state ###
Mm = np.zeros(len(qmmm))
Mm[0] = 1
Mm[12] = 1
qmmm.set_initial_magnetic_moments(Mm)

### Create friction array ###
# Since production run is NVE, it is just a bunch of zeros
fc = np.zeros((len(qmmm),1))
fc[qmidx::] = 0.0

### Create QM cell ###
pos = qmmm[:qmidx].get_positions()
rcut = 4.0
qmC = np.zeros((3,3))
xmin = pos[:,0].min(); xmax = pos[:,0].max()
qmC[0,0] += xmax - xmin + 2 * rcut
ymin = pos[:,1].min(); ymax = pos[:,1].max()
qmC[1,1] += ymax - ymin + 2 * rcut
zmin = pos[:,2].min(); zmax = pos[:,2].max()
qmC[2,2] += zmax - zmin + 2 * rcut

### Set up GPAW calculator ###
# MixerDif is best for spin-unrestricted calculations:
mixer = MixerDif(beta=.15, nmaxold = 5, weight=10, beta_m=0.15, nmaxold_m=5,
    weight_m=10.0)
```

```
calc = GPAW(
    gpts=(112,120,112),
    mode='lcao',
    basis={None:'dzp', 'Ir':'tzp'},
    xc = 'PBE',
    txt = PATH+outname+'.out',
    poissonsolver=PS(relax='GS', eps=1.0e-7),
    stencils=(3,3),
    charge=2,
    mixer=mixer,
    occupations=FD(width=.05, fixmagmom=True),
    convergence={'density':2.5e-4},
    maxiter=200,
    nbands=-80

### Attach QM/MM calculator ###
# calc_1: Takes the QM-MM interface calculator. Inputs: The MM LJ array, and
    number of atoms per solvent molecule (so as to not break up molecules under
    PBC operations)
# calc_2: GPAW, the cutoff, LJ arrays, and the QM cell.
qmmm.set_calculator(
        ase_qmmm(qmmm, index=qmidx,
        calc_1 = Inter_Pot(LJ_mm[:,:3],3),
        calc_2 = calc, rcut = rcut, LJ_mm = LJ_mm, LJ_qm = LJ_qm, cell = qmC))

### make a dynamics object ###
# This langevin dynamics version is modified to work with the QM/MM code
dyn = LC(qmmm, 1.0*unit.fs, unit.kB*300., fc, index=qmidx, sol_index=3, list_sub
    =qm_con, list_sol=[(0,1,1.458),(0,2,2.615),(1,2,1.157)])
# sol_index: Amount of atoms pr solvent molecule
# list_sub: Constraints of QM subsystem
# list_sol: Constraints in MM subsystem

### Attach trajectory and run ###
traj = PT(PATH+outname+'.traj', 'w', qmmm)
dyn.attach(traj.write,interval = 1)
dyn.run(<number of steps to run>)
```

Listing 3: Script for getting parameters out of trajectory files

```
#!/usr/bin/env python
import subprocess as sp
from subprocess import PIPE
#from subprocess import Popen, PIPE
import numpy as np
import os
from sys import argv
from ase.io.trajectory import PickleTrajectory as PT
"""
Script that uses the output functionality of ag to write
parameters from one or more trajectories to output files
with <trajectoryname>.dat filenames.
The last column will contain a cumulative time over all
trajectories. The input format is as follows:
./GetData.py <trajname>.traj@::<timestep>
where <timestep> is the time specified in the input script.
e.g:
./GetData.py traj1.traj@::0.5 traj2.traj@::2 ...

                                        Asmus O. Dohn 2014
"""

# remove script name from argv, the rest is trajs and props:
del argv[0]
trajs = []
props = []
tsteps = []


for filename in argv:
    if ".traj" in filename:
        if filename[-5:] == ".traj":
            n = "1"
            trajs.append(filename)
            tsteps.append(n)
        elif "@" in filename:
            n = filename.split(":")[-1]
            filetrunk = filename.split('@::')[0]
            trajs.append(filetrunk)
            tsteps.append(n)
        else:
            print "bad traj input!"
            break
```

```
props = ['d(0,1)',    #Distance between atom 0 and 1
                'a(56,1,74)',#Angle bewtween atms 56,1,74
                'dih(68,73,79,74)', #Dihedral
                'fave', # Total average force
                'fmax', # Maximum force
                'np.sqrt(F[40,0]**2+F[40,1]**2+F[40,2]**2)', # Magnitude of
                    force on atm 40
              ]

propnames = [] # list of what you want the labels of the props above to be in
    the headerline

def getdata(traj,prop):
    #print prop
    out = sp.Popen(["ag", "-t",  "-g", prop, traj], stdout=PIPE)
    out = out.stdout.read()
    out = map(float, out.strip().split('\n')) #remove last \n with strip and
    out = np.array(out)
    return out

def getrealtime(traj,ts,prevstep):
    ts = float(ts)
    thistraj = PT(traj)
    tl = len(thistraj)    #from     to             numpoints
    realtime = np.linspace(prevstep,prevstep+tl*ts,tl,endpoint=False)
    print str(realtime)
    return realtime


# define function that gets the trajectories
def gettraj(path):
    files = []
    dir_list = os.listdir(path)
    name_to_check_for = '.traj'
    for i in range(len(dir_list)):
        if dir_list[i].endswith(name_to_check_for):
            files.extend([dir_list[i]])
    return files
```

```
if len(trajs) is 0:
    print "no files specified looking for .trajs in current directory"
    path = os.curdir+'/'
    trajs = gettraj(path)
    trajs.sort()
    for i in traj:
        print i

# get the data
ct = 0
for traj in trajs:
    print traj+', tstep = '+tsteps[ct]+' ...'
    data = []
    trajtrunk = os.path.splitext(traj)[0]
    ofname = trajtrunk+'.dat'
    f = open(ofname, 'w')
    for prop in props:
        #print prop
        thispath = os.curdir+'/'+traj
        thisdata = getdata(thispath,prop)
        data.append(thisdata)
    # Make last vector of realtime
    if ct == 0:
        prevstep = 0
        realtime = getrealtime(traj,float(tsteps[ct]),prevstep)
        data.append(realtime)
    else:
        prevstep = realtime[-1]+float(tsteps[ct])
        realtime = getrealtime(traj,float(tsteps[ct]),prevstep)
        data.append(realtime)
    # Write data to file
    f.write(''.join(format(x, 's').ljust(16) for x in propnames))
    f.write('\n')
    transdata = map(list, zip(*data))
    for i in range(len(transdata)):
        f.write(''.join(format(i, '2.8f').ljust(16) for i in transdata[i]))
        f.write('\n')
    f.close()
    ct +=1
```

Listing 4: Script for exporting trajectories with the QM solute centered in the solvent box, in the XYZ format. For e.g. RDF calculations in VMD. - Or visualization. The centering using the minimum image convention is based upon how it is done internally in the QM/MM code

```
#!/usr/bin/env python
import subprocess as sp
from subprocess import PIPE
import numpy as np
import os
from sys import argv
from ase.io import read, write
from ase.visualize import view
from ase.io.trajectory import PickleTrajectory as PT

def micwrap(qmmm,qmidx):
    mp = 3 #define Particles in solvent Mol
    oldcell = qmmm.get_cell()
    # Create QM cell
    pos = qmmm[:qmidx].get_positions()
    oldpos = pos
    rcut = 4.0
    xmin = pos[:,0].min(); xmax = pos[:,0].max()
    ymin = pos[:,1].min(); ymax = pos[:,1].max()
    zmin = pos[:,2].min(); zmax = pos[:,2].max()

    qmC = np.zeros((3,3))
    qmC[0,0] += xmax - xmin + 2 * rcut
    qmC[1,1] += ymax - ymin + 2 * rcut
    qmC[2,2] += zmax - zmin + 2 * rcut
    ### Centering IrD in the QM cell and the solvent around it
    # for the fixing to work
    qm_subsystem = qmmm[:qmidx]
    mm_subsystem = qmmm[qmidx:]
    qm_subsystem.set_cell(qmC)
    qm_subsystem.center()
    newpos = qm_subsystem.get_positions()
    trans = oldpos - newpos
    transvec = trans[0,:]
    mm_newpos = mm_subsystem.get_positions()
    mm_newpos -= transvec
    qmmm.set_cell(qmC)
    mm_subsystem.set_positions(mm_newpos)
    newqmmm = qm_subsystem + mm_subsystem
    ## and now MIC
    index = qmidx
    pos = newqmmm[index:].get_positions()
    # Minimum image relative to the center of the qm cell!
    n = np.zeros(np.shape(pos))
```

```
    c_mid = qm_subsystem.cell.diagonal() * 0.5
    n[::mp] = np.rint((c_mid - pos[::mp]) / oldcell.diagonal())
    # Grab all atoms of this particular molecule
    for i in range(1,mp):
        n[i::mp] += n[::mp]

    pos += n * oldcell.diagonal()
    mm_subsystem = newqmmm[index:]
    mm_subsystem.set_positions(pos)
    mm_subsystem.set_pbc((1,1,1))
    finalqmmm = qm_subsystem+mm_subsystem
    finalqmmm.set_cell(oldcell)
    finalqmmm.center()
    return finalqmmm


def xyzwrite(thistraj):
    traj = PT(thistraj)
    tl = len(traj)
    numatms = len(traj[0])

    oname = thistraj[0:-4]
    print 'Writing '+oname+'xyz...'
    f = open(oname+'xyz','w')
    f.write('%d\n' % numatms)
    f.write('Stuff about the simulation. Step 1\n')
    for i in range(tl):
        wstep = micwrap(traj[i],130) #wrap it up - remember to change to your
            qmidx!!
        ThisPos = wstep.get_positions()
        ThisSym = wstep.get_chemical_symbols()
        if i is not 0:
            f.write('%d\n' % numatms)
            f.write('Step: %d\n' %(i+1))
        for j in range(numatms):
            f.write('%3s%14.6f%14.6f%14.6f\n' % (ThisSym[j],ThisPos[j,0],ThisPos
                [j,1],ThisPos[j,2]))
    f.close()

# Input fixing - specify trajs as with GetData scripts...
del argv[0]
trajs = []
for i in argv:
    if i.endswith('.traj'):
        trajs.append(i)

for thistraj in trajs:
    xyzwrite(thistraj)
```

Matlab Parsers For ASE/GPAW Output

Listing 5: Matlab .out file parser

```
function [DataCell, Data, Output] = rdGPAWout(DataFile)
fileID = fopen(DataFile,'r');

InFile = 1;

StepCt = 0;
CoCStepCt = 0;
SCStepCt = 0;
LMMCt = 0;
TotECt = 0;
QMCCt = 0;
PosCt = 0;
cStepIsConverged = 0;
pStepIsConverged = 0;
while InFile
    Line=fgetl(fileID);
    if Line == -1
        InFile = 0;
        break
    end
    if strfind(Line,'Energy Contributions Relative to Reference Atoms:')
        Line=fgetl(fileID); % get over the ----
        InBlock = 1;
        TotECt = TotECt + 1;
        while InBlock
            Line=fgetl(fileID);
            spLine = strsplit(Line,':');
            if strfind(spLine{1},'Entropy')
                Ene(TotECt).Entropy = str2num(spLine{2});
            elseif size(spLine,2) == 2;
                Ene(TotECt).(strrep(spLine{1}, ' ', '_')) = str2num(spLine{2});
            elseif isempty(Line) % empty line efter Zero Kelvin
                InBlock = 0;
                break
            end
        end
        cStepIsConverged = 1;
        pStepIsConverged = 1;
    end
    if ~isempty(strfind(Line,'Unit Cell:')) && cStepIsConverged
        QMCCt = QMCCt +1;
        CellNames=fgetl(fileID); % Periodic     X          Y          Z
            Points  Spacing
        Line=fgetl(fileID);
        CellNames = strsplit(CellNames,' ');
        for ii = 1:3
            Line=fgetl(fileID);
            spLine = strsplit(Line,' ');
            for jj = 1:size(CellNames,2)
                QMCArray{ii,jj} =spLine{jj+2};
            end
```

```
        end
        QMCell{QMCCt} = [CellNames; QMCArray];
        cStepIsConverged = 0;
    end
    if ~isempty(strfind(Line,'Positions:')) && pStepIsConverged
        PosCt = PosCt + 1;
        InBlock = 1;
        BlockCt = 0;
        while InBlock
            Line=fgetl(fileID);
            if ~isempty(Line)
                BlockCt = BlockCt + 1;
                spLine = strsplit(Line,' ');
                Elements{BlockCt} = spLine{2};
                X(BlockCt) = str2num(spLine{3});
                Y(BlockCt) = str2num(spLine{4});
                Z(BlockCt) = str2num(spLine{5});
            else
                InBlock = 0;
                break
            end
        end
        Output.Pos(:,:,PosCt) = [X' Y' Z'];
        pStepIsConverged = 0;
    end
    if strfind(Line, 'Fermi Levels:') % get Fermi Levels and check if restricted
         or unrestricted
        FLLine = strsplit(Line,'Fermi Levels:');
        FLLine = FLLine{2};
        FLLine = strsplit(FLLine,',');
        FL1 = str2num(FLLine{1});
        FL2 = str2num(FLLine{2});
        if FL1-FL2 == 0
            SpinRestricted = 1;
        else
            SpinRestricted = 0;
        end
    end
    if strfind(Line, 'Center of Charge:')
        CoCStepCt = CoCStepCt +1;
        CoC = strsplit(Line, 'Center of Charge: ');
        CoC = strsplit(CoC{2},' ]');
        CoC = strsplit(CoC{1},'[ ');
        CoC = str2num(CoC{2});
        CoCs(CoCStepCt,:) = CoC;
    end
```

```
if strfind(Line, 'Spin contamination:')
    SCStepCt = SCStepCt +1;
    SC =strsplit(Line,'Spin contamination: ');
    SC = strsplit(SC{2}, ' electrons');
    SCs(SCStepCt) = str2num(SC{1});
end
if ~isempty(strfind(Line,'Occupancy')) && SpinRestricted
    StepCt = StepCt + 1;
    InBlock = 1;
    BlockLineCt = 0;
    FL(StepCt) = FL1;
    while InBlock
        Line = fgetl(fileID);
        NumLine = str2num(Line);
        if isnumeric(NumLine) && ~isempty(NumLine)
            BlockLineCt = BlockLineCt + 1;
            DataCell{StepCt}(BlockLineCt,:) = NumLine;
        else
            InBlock = 0;
            break
        end
    end
    if mod(StepCt,100) == 0;
        display(['Finished Reading SpinRestricted Calc. Step: ' num2str(
            StepCt)])
    end
elseif ~isempty(strfind(Line,'Occupancy')) && ~SpinRestricted % Go into orb.
     energies
    StepCt = StepCt + 1;
    InBlock = 1;
    BlockLineCt = 0;
    FL(StepCt,1) = FL1;
    FL(StepCt,2) = FL2;
    while InBlock % actually works the exact same way lol. Only diff is
        having the FL(StepCt,2)
        Line = fgetl(fileID);
        NumLine = str2num(Line);
        if isnumeric(NumLine) && ~isempty(NumLine)
            BlockLineCt = BlockLineCt + 1;
            DataCell{StepCt}(BlockLineCt,:) = NumLine;
        else
            InBlock = 0;
            break
        end
    end
    if mod(StepCt,100) == 0;
        display(['Finished Reading SpinUnRestricted Calc. Step: ' num2str(
            StepCt)])
    end
```

```
    elseif ~isempty(strfind(Line,'Local Magnetic Moments:')) && ~SpinRestricted
        LMMCt = LMMCt +1;
        InBlock = 1;
        BlockLineCt = 0;
        while InBlock
            Line = fgetl(fileID);
            NumLine = str2num(Line);
            if isnumeric(NumLine) && ~isempty(NumLine)
                BlockLineCt = BlockLineCt + 1;
                LMMCell{LMMCt}(BlockLineCt,:) = NumLine;
            else
                InBlock = 0;
                break
            end
        end
        if mod(LMMCt,100) == 0;
            display(['Finished Reading SpinUnRestricted LMM. Step: ' num2str(
                LMMCt)])
        end

    end
end

% Make nicer array
StepLengths = zeros(size(DataCell,2),1);
for ii = 1:length(StepLengths)
    StepLengths(ii) = size(DataCell{ii},1);
end
MaxLength = max(StepLengths);
Data = NaN(MaxLength,size(DataCell{1},2),size(DataCell,2));
for ii = 1:size(DataCell,2);
    Data(1:size(DataCell{ii},1),1:size(DataCell{ii},2),ii) = DataCell{ii};
end

if exist('LMMCell')
    StepLengths = zeros(size(LMMCell,2),1);
    for ii = 1:length(StepLengths)
        StepLengths(ii) = size(LMMCell{ii},1);
    end
    MaxLength = max(StepLengths);
    LMMData = NaN(MaxLength,size(LMMCell{1},2),size(LMMCell,2));
    for ii = 1:size(LMMCell,2);
        LMMData(1:size(LMMCell{ii},1),1:size(LMMCell{ii},2),ii) = LMMCell{ii};
    end
end
```

```
Output.Energies = Data;
Output.FermiLevels = FL;
if exist('CoC')
  Output.CoCs = CoCs;
end
if exist('SCs')
   Output.SCs  = SCs;
end
if exist('LMMData')
  Output.LMM = LMMData;
end
Output.E = Ene;
Output.Cell = QMCell;
Output.Elements = Elements;
end
```

B.2 Miscellaneous

Listing 6: Script for VMD to calculate an RDF from each frame, which can then later be re-binned in e.g. Matlab. Made for the Iridium dimer study. This version calculates the Ir–N pairwise RDF

```
cd #<dir with trajectories, here xyz files, since VMD does not read ase .traj>
proc dirlist { dir ext } {
    set contents [glob -directory $dir *$ext]
    foreach item $contents {
       append out $item
       append out "\n"
       }
    return $out
    }

set runtypes {Down Top Up Well}

foreach runtype $runtypes {
        set xyzs [dirlist RDFs/$runtype/ .xyz]
        set ct 0
        foreach file $xyzs {
           incr ct 1
           mol load xyz $file
           set trajlen [molinfo top get numframes]
           for {set i 0} {$i < $trajlen} {incr i} {
                 #set rdf $rdfs($i)
                 #set atm1 $atms1($i)
                 #set atm2 $atms2($i)
```

```
                puts "Making RDF for atmselect Ir and solvent -> $file\_$i\
                    .dat"
                set outfile1 [open RDFs/DATFILES_IrSolvN/$runtype/$ct/$i\.dat
                    w]
                molinfo top set frame $i
                molinfo top set a 28; # Set cell dims, since they are not
                    saved in the xyz format
                molinfo top set b 28.5
                molinfo top set c 31.5
                molinfo top set alpha 90
                molinfo top set beta 90
                molinfo top set gamma 90
                set sel1 [atomselect top "name Ir"]
                set sel2 [atomselect top "name N and not residue 0"]
                set gr0 [measure gofr $sel1 $sel2 delta 0.2 rmax 15.0 usepbc 1
                     selupdate 0 first $i last $i step 1]
                set r [lindex $gr0 0]
                set gr [lindex $gr0 1]
                set igr [lindex $gr0 2]
                set isto [lindex $gr0 3]
                foreach j $r k $gr l $igr m $isto {
                      puts $outfile1 [format "%.4f\t%.4f\t%.4f\t%.4f" $j $k $l
                          $m]
                }
                close $outfile1
        }
        mol delete top
    }
}
```

Listing 7: Tcl script for calculating the solute-solvent pairwise RDFs, averaged over frames from `tstart` to the last frame. The code is easily changed to produce *all* pairwise RDFs for e.g. X-ray scattering simulations, by expanding the `atms` and `ratms`-arrays

```
cd #<OutputDir>
set tstart 0

set Ru "index 0"
set Co "index 1"
set NSolute "name N and index < 124"
set CSolute "name C and index < 124"
set HSolute "name H and index < 124"
set CSolvent "name C and index > 124"
set NSolvent "name N and index > 124"
set HSolvent "name H and index > 124"
set MeSolvent "name C and index > 124 and (not within 1.5 of (name N and index >
    124))"
set CmSolvent "name C and index > 124 and (within 1.5 of (name N and index >
    124))"


set rRu "Ru"
set rCo "Co"
set rNSolute "NSolute"
set rCSolute "CSolute"
set rH "HSolute"
set rCSolvent "C"
set rNSolvent "N"
set rHSolvent "HSolvent"
set rMeSolvent "Me"
set rCmSolvent "C"

array set atms1 [list 1 $Ru 2 $Co 3 $NSolvent 4 $MeSolvent 5 $CmSolvent]
array set atms2 [list 1 $Ru 2 $Co 3 $NSolvent 4 $MeSolvent 5 $CmSolvent]
array set ratms1 [list 1 $rRu 2 $rCo 3 $rNSolvent 4 $rMeSolvent 5 $rCmSolvent]
array set ratms2 [list 1 $rRu 2 $rCo 3 $rNSolvent 4 $rMeSolvent 5 $rCmSolvent]

array set runs [list 0 "GS" 1 "LS" 2 "HS"]
for {set ii 0} {$ii < 3} {incr ii} {
   set name $runs($ii)
   mol top $ii
```

```
    pbc set {32.0 32.0 44.0} -all
    cd $name
    for {set i 1} {$i < 6} {incr i} {
       for {set h $i} {$h < 6} {incr h} {
          if { $i==1 && $h==1} {
             continue
             } elseif { $i==2 && $h==2} {
             continue
             } else {
          set rdf $ratms1($i)-$ratms2($h)
          set atm1 $atms1($i)
          set atm2 $atms2($h)
          puts "Making RDF for atmselect $atm1 & $atm2 -> $name\_$rdf\.dat"
          set outfile1 [open $name\_$rdf\.dat w]
          set sel1 [atomselect top $atm1]
          set sel2 [atomselect top $atm2]
          set gr0 [measure gofr $sel1 $sel2 delta 0.2 rmax 16.0 usepbc 1
              selupdate 0 first $tstart last -1 step 1]
          set r [lindex $gr0 0]
          set gr [lindex $gr0 1]
          set igr [lindex $gr0 2]
          set isto [lindex $gr0 3]
          foreach j $r k $gr l $igr m $isto {
            puts $outfile1 [format "%.4f\t%.4f\t%.4f\t%.4f" $j $k $l $m]
          }
          close $outfile1
             }
       }
    }
    cd ..
}
```

MIX
Papier aus verantwortungsvollen Quellen
Paper from responsible sources
FSC® C105338

If you have any concerns about our products,
you can contact us on
ProductSafety@springernature.com

In case Publisher is established outside the EU,
the EU authorized representative is:
Springer Nature Customer Service Center GmbH
Europaplatz 3, 69115 Heidelberg, Germany

Printed by Libri Plureos GmbH
in Hamburg, Germany